AN INTRODUCTION TO
SOIL
SCIENCE

by

E. A. FITZPATRICK

D.I.C.T.A., Ph.D.

Department of Soil Science
The University of Aberdeen

OLIVER & BOYD

EDINBURGH

OLIVER AND BOYD LTD

Croythorn House
23 Ravelston Terrace, Edinburgh, EH4 3TJ

A Division of Longman Group Limited

ISBN 0 05 002777 8

Printed in Hong Kong by
Wah Cheong Printing Press Ltd.

Contents

Colour plates

Acknowledgements

I should like to thank those persons who have helped in the preparation of this book through their suggestions or by providing material in one form or another. To Professor J. Tinsley, Dr. J. R. H. Coutts, Dr. D. A. MacLeod, Dr. C. E. Mullins and Dr. J. W. Parsons for reading and commenting on the material at various stages.

To Mr. I. G. Pirie for helping to prepare Figs. 1, 6, 9, 11, 12, 29, 30, 57 and to the following for permission to use various illustrations: The British Museum for Fig. 8; F. M. Synge for Fig. 25; USDA Soil Conservation Service for Figs. 24, 39, 49, 54; Aerofilms Ltd. for Figs. 22, 53, part of 66; Dairy Science Dept., Univ. of Florida For Fig. 48; Paul Popper Ltd. for Fig. 55; Massey-Ferguson for Fig. 56; Soil Survey of England and Wales for Fig. 57; Ordnance Survey for Fig. 59; HMSO for Fig. 64; FAO for part of Fig. 66, and the University of Aberdeen for Figs. 16, 36, 37, 38, and 42.

Introduction

After many centuries of the use and misuse of soils there is now a rapidly growing consciousness about the place of soils in the environment and their importance as vital factors in the life of most organisms. Perhaps it is true to say that soils are man's major natural resource because most of our food and clothing comes directly or indirectly from them. Since most soils take thousands or even millions of years to form they cannot be replaced if they are washed away by erosion. It is therefore of paramount importance that our soil mantle be carefully nurtured so that it will be preserved in a healthy and fertile state for generation after generation.

This book is intended for beginners who would like to know something about the nature and formation of soils and also about their uses and geographical variation over the surface of the earth. Pupils in their final years at school should find this book useful, and it should also be helpful to the large number of biology, geography and geology students in their first year at universities and other institutes of higher learning. There are also a number of generally interested people for whom this should provide the breadth and depth of information they require.

Soils, or the pedosphere, are composed of air – water – mineral material – organic material and organisms, and can be regarded as an amalgam of the lithosphere, the biosphere, the hydrosphere and the atmosphere as shown diagrammatically in Fig. 1. The amount and type of these constituents varies widely from place to place over the earth's surface causing an almost infinite variability in the types of soils. Such a wide range induces many complications particularly with regard to classifying soils. Thus at present there are no less than ten different systems in use in different parts of the world. Therefore in order to make this book of appeal to the widest public, very little technical terminology has been used, apart from those terms that have a measure of international recognition. On the other hand the wide variability in soils presents a challenge and holds an evergreen fascination for the soil scientist.

Soils can be studied in a variety of ways. They can be considered as natural phenomena and worthy of independent study or they can be studied in relation to the natural environment. Probably the greatest amount of study is directed towards establishing the distribution and mapping of soils, and determining their suitability for crop production, be they food crops or trees.

At the end of the text there are a number of Appendices containing particular information about practical exercises and projects that may be conducted in order to gain a fuller appreciation of the properties of soils.

In addition there is a glossary containing a number of the common terms used in soil science.

At present great emphasis is being placed on the utilisation of soils but before they can be utilised properly they must be recognised as highly organised physical, chemical and biological systems whose nature, properties, and place in the environment should be understood.

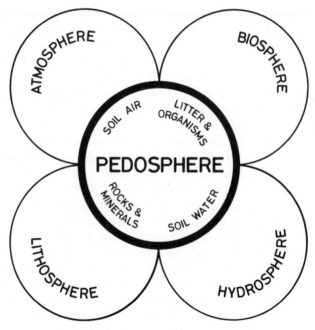

FIG. 1. The pedosphere

1

Fundamental concepts

Most people usually think of soils as the upper few centimetres of the earth's crust permeated by plant roots or cultivated. This is a somewhat limited approach which focuses attention mainly on soils as media for plant growth. In the first instance soils should be regarded as natural phenomena and part of the environment.

Soils as natural phenomena

The soil scientist recognises not only the top soil in which plants grow but also many other layers beneath, and his first step towards understanding soils is to dig a pit into the surface of the earth and to carry out visual observations. The depth of the pit is determined by the nature of the soil itself and normally varies from one to three metres, below which is relatively unaltered material.

The pit reveals a characteristically layered pattern mainly expressed through differences in the colour of the individual layers. Each layer is known as a *horizon* and the set of layers in a single pit is termed a *soil profile*. The two examples given in Figs. 2 and 3 illustrate the main principles concerned with soil profiles as well as some of the variability that exists between soils found under certain types of natural coniferous and broad-leaved deciduous forest.

Fig. 2 and Colour Plate IB show a horizon sequence in a podzol profile – a soil of common occurrence in humid temperate areas under coniferous forest. At the surface there is an accumulation of freshly fallen plant litter, below which is dark brown, partially decomposed plant material, mainly inhabited by fungi and small arthropods which are largely responsible for its break down. This grades into very dark brown or black amorphous organic matter in an advanced stage of decomposition. Beneath the organic layers there is a very dark grey mixture of black organic matter and light coloured bleached mineral grains, mainly quartz. This horizon is underlain by a pale grey or white horizon composed mainly of bleached

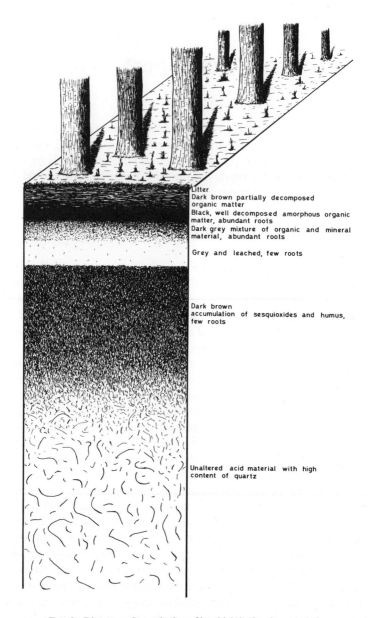

Litter
Dark brown partially decomposed organic matter
Black, well decomposed amorphous organic matter, abundant roots
Dark grey mixture of organic and mineral material, abundant roots

Grey and leached, few roots

Dark brown
accumulation of sesquioxides and humus, few roots

Unaltered acid material with high content of quartz

FIG. 2. Diagram of a podzol profile which is the characteristic soil of the northern coniferous forests

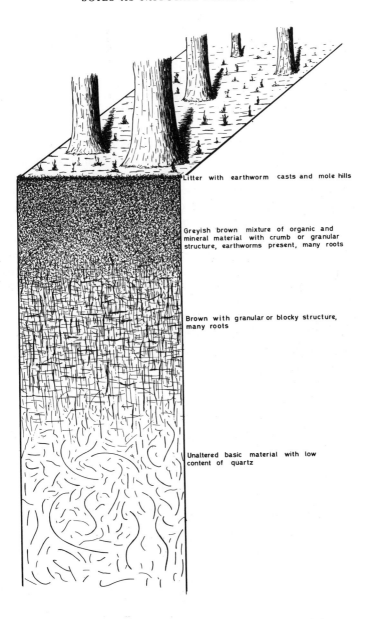

Litter with earthworm casts and mole hills

Greyish brown mixture of organic and mineral material with crumb or granular structure, earthworms present, many roots

Brown with granular or blocky structure, many roots

Unaltered basic material with low content of quartz

FIG. 3. Diagram of an altosol profile which is the characteristic soil of the temperate deciduous forests

quartz. The mineral material in these two horizons is strongly weathered and percolating water removes the colouring substances which are mainly compounds of iron. The substances removed from the upper horizons accumulate immediately below to form a middle horizon with its characteristic brown or dark brown colour. Finally there is the relatively unaltered material which is usually very sandy containing a high proportion of quartz and is therefore acid in composition.

Fig. 3 and colour Plate IIIA are examples of altosol (brown earth) profiles that develop under deciduous forests and from basic material which is characterised by having minerals containing large amounts of basic cations. At the surface there is a thin loose litter of leaves and twigs containing numerous earthworm casts and mole hills. Beneath is a brown or greyish brown horizon of organic and mineral material, intimately mixed mainly by earthworms. This horizon grades into a middle brown horizon showing less evidence of faunal activity, followed by relatively unaltered material. This soil has not the marked contrast between horizons as shown by the podzol because in this case iron is not translocated from the upper to the middle horizon.

The letters A, B and C are usually used to designate the upper, middle and lower horizons respectively, but these letters have different morphological and genetic connotations to different workers and their usage can at times be very misleading.

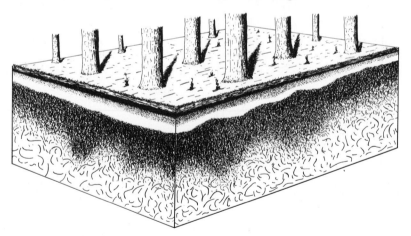

FIG. 4. A three dimensional block of soil

Both podzols and altosols are cultivated in a variety of ways leading to marked changes in the character of the upper horizons. Podzols are naturally infertile and need a considerable amount of treatment before agricultural crops can be grown successfully. On the other hand altosols have a high natural fertility because of the basic material and vigorous soil fauna and flora.

The soil profile is simply a two dimensional section through soil which in reality extends laterally in all directions over the surface of the earth forming a three-dimensional continuum (Fig. 4). Further, the constituent horizons do not remain uniform throughout their lateral extent but exhibit a gradual change from one set of horizons to another.

A very important aspect of soils is that they seldom if ever have sharp boundaries, usually they grade gradually from one into another. This is often overlooked when isolated profiles in separate pits are examined. Fig 5 attempts to illustrate how soils change from one to another when travelling across the landscape. At stage one there are two horizons – solid circles and solid hexagons, while solid triangles represent the parent material for the whole sequence. At stage two the solid circles and solid hexagons have changed into mixtures of solid and open circles overlying solid and open hexagons and at stage three there are only open circles overlying open hexagons. These changes represent the situation in which the whole soil is changing gradually on travelling across the landscape. Stages four and five represent the situation in which the upper horizon remains uniform but the middle gradually changes from open hexagons to solid rectangles. Thus, not only do soils change from place to place over the landscape but the nature of the change can also vary.

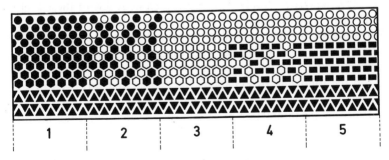

FIG. 5. Intergrading of horizons

Finally, it should be stated that soils are not static but are continually forming and changing so that they form a natural space-time continuum.

Soils as part of the environment

Soils can be considered as a product of the environment – the attitude taken by the pure soil scientist – or they can be regarded as a part of the environment, the more general attitude taken by the natural scientist. When the latter attitude is adopted, soils are then part of a larger and highly complex system which for convenience can be broken down into a number of simpler cycles and relationships, the most important including:

> The carbon cycle (page 42)
> The moisture cycle (page 18)
> The nitrogen cycle (page 44)
> Energy relationships (page 13)
> The oxygen cycle
> The mineral cycle

Prior to the last century the influence of man on these cycles could be regarded as relatively small but since the industrial revolution he has increasingly influenced his environment and in some cases exercised almost absolute control. Perhaps the classical example is through the increased growth of crops by the use of fertilizers and pesticides resulting in a general benefit to mankind, but the application of too many fertilizers, insecticides, or weed killers can lead to their presence in drainage waters which ultimately will pollute rivers and lakes thus negating the benefits from an increased food yield, because the water supply is rendered useless. A few of these cycles will be discussed in Chapters 3 and 5 from the standpoint of minimum disturbance by man as well as mentioning some of the dramatic disruption that can take place when unforseen changes in the soil are introduced.

In the following two chapters the factors and processes of soil formation are considered and together they illustrate that soils are formed by the interaction of certain specific environmental factors.

2

Factors of soil formation

Dokuchaev, the famous Russian soil scientist, showed that soils do not occur by chance but they usually form a pattern in the landscape, and furthermore he firmly established that they develop as a result of the interplay of five factors: parent material, climate, organisms, topography and time (Fig. 6). Below are presented the more important characteristics of each factor.

Parent material

Jenny (1941) defines parent material as "the initial state of the soil system". The precision of this definition cannot be questioned but

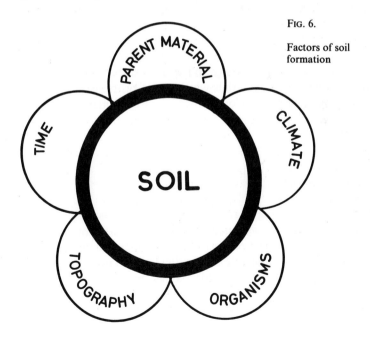

FIG. 6.

Factors of soil formation

it is often difficult to determine the initial stages of soils for in a number of cases the original character of the material has been changed markedly by a long period of soil formation.

Often the unaltered material in the lower part of the profile is similar to the material from which the horizons above have formed but this is not always the case. For example, there may be a thin cover of drift overlying rock as shown in Plate IIIA. In fact recent studies have shown that many if not most soils display some evidence of layering hence the difficulties in applying the A, B, C designation of soil horizons.

Parent materials are made up of mineral material or organic matter or a mixture of both. The mineral material which is the most widespread type of parent material contains a large number of different rock-forming minerals and can be in either a consolidated or an unconsolidated state. The consolidated mineral material includes rocks like granite, basalt, schist and conglomerate and the unconsolidated material comprises a wide range of superficial deposits of which glacial drift (Fig. 21), alluvium and loess (Fig. 23), are three important representatives. The most important properties of mineral parent materials are their chemical and mineralogical properties; these are responsible largely for the course of soil formation and the resulting chemical mineralogical and physical composition of the soil, including the secondary minerals produced by weathering. Mineral parent materials also contribute to soil formation through their permeability and specific surface area.

The organic parent materials which are of restricted distribution are usually composed predominantly of unconsolidated dead and decaying plant remains.

Chemical and mineralogical composition
of parent materials

The minerals occurring in rock such as granite, basalt, gneiss and schist are termed *primary minerals*. In the soil these minerals are decomposed to form *secondary minerals*, particularly the clay minerals. Generally, minerals can be divided into non-silicates and silicates. The non-silicates include oxides, hydroxides, sulphates, chlorides, carbonates and phosphates. Most have relatively simple structures but they vary widely in their solubility and resistance to decomposition. The silicate minerals have very complex structures in which the fundamental unit is the silicon-oxygen tetrahedron. This is composed of a central silicon ion surrounded by four closely packed and equally spaced oxygen ions. The whole forms a pyramidal structure, the base of which is composed of three oxygen ions with the fourth forming the apex (Fig. 7). The tetrahedra themselves are linked together in a number of different ways forming silicates with a variety of distinctive and characteristic patterns which form the basis of the classifications of these minerals. An important variation in the tetrahedral structure is the substitution of Al for Si; this is known as *isomorphous replacement* and causes an imbalance in the charges which is satisfied by cations such as Na, K, Ca, Mg and Fe. Another important unit is the aluminium-hydroxyl octahedron which is composed of an aluminium ion surrounded by six closely packed hydroxyl groups. Like the silicon-oxygen tetrahedra, aluminium hydroxyl octahedra form sheets known as gibbsite sheets which occur within the structure of many sheet silicates (see below).

There are four main groups of silicates; the framework silicates, the chain silicates, the ortho- and ring-silicates, and the sheet silicates. The framework silicates include quartz and the feldspars, the chain silicates include the pyroxenes and amphiboles, the ortho- and ring silicates vary widely in their composition and include the olivines and zircon. The sheet silicates include the micas (biotite and muscovite) and the all important clay minerals.

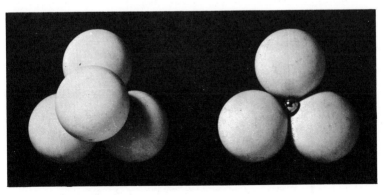

FIG. 7. Models of silicon-oxygen tetrahedra (*left*) complete model, (*right*) model with apical oxygen ion removed to show the smaller silicon ion

Quartz is one of the most common minerals in soils and is composed of silicon-oxygen tetrahedra linked through the oxygen ions, therefore it contains only silicon and oxygen. In rocks it is seen as colourless or milky crystals with concoidal fractures and is among the minerals most resistant to weathering, accumulating as one of the main residues during weathering.

Feldspars are white, pale grey or pale pink minerals that are very common in most igneous and metamorphic rocks. They are composed largely of silicon and oxygen but they also contain large amounts of aluminium due to isomorphous replacement, with varying amounts of sodium, potassium and calcium satisfying the extra charge.

The pyroxenes are generally colourless to dark green silicates, occurring principally in basic igneous rocks. In addition to silicon and oxygen they contain large but varying amounts of iron, calcium and magnesium.

The amphiboles are characterised by hornblende which is a green silicate and occurs in many igneous and metamorphic rocks. This mineral contains large amounts of iron, magnesium and calcium.

The principal sheet silicates occurring in igneous and metamorphic rocks are brown biotite and colourless muscovite. Most of the clay minerals are also sheet silicates with kaolinite being the simplest type. Kaolinite is composed of compound layers, each of which is composed of a silicon tetrahedral sheet on either side of an aluminium octahedral or gibbsite sheet (Fig. 8).

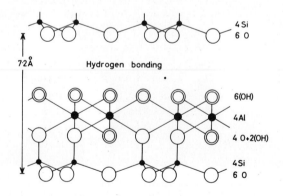

FIG. 8. Structure of kaolinite

The upper 10 km of the earth's crust is made up predominantly of the following elements:

Element	Per cent (by wt.)	Element	Per cent (by wt.)
Aluminium	8	Oxygen	47
Calcium	4	Phosphorus	< 1
Carbon	< 1	Potassium	3
Chlorine	< 1	Silicon	28
Iron	5	Sodium	3
Magnesium	2	Sulphur	< 1
Manganese	< 1	Titanium	< 1

Oxygen occupies 90 per cent by volume and is the most abundant element but it usually occurs in combination with other elements, particularly with silicon mainly within the more complex silicate structures. Thus the majority of the resulting soils are composed predominantly of silica, either as quartz or in a combined form in silicates. The third most frequent element is aluminium which is found mainly in feldspars and in sheet silicates but which occurs also in varying proportions in the other silicates. Iron is also of fairly widespread distribution, occurring in large amounts in relatively few minerals of which biotite, some pyroxenes, amphiboles and olivines are the principal contributors and, it should be noted that it is mainly in the ferrous state. Sodium and potassium are found chiefly in the feldspars but significant amounts of potassium also occur in the micas. Calcium and magnesium have a wide distribution among the silicates as well as being the principal cations in many non-silicates such as calcite and dolomite which are the dominant minerals in limestone. Phosphorus is very restricted in its distribution occurring only in significant proportions in apatite which is fairly widespread in small amounts in parent materials but the total content of this mineral in soils seldom exceeds 0·2 per cent. The microelements needed by plants (see page 81) occur in a number of minerals principally those of low frequency such as tourmaline, zircon and rutile which are usually referred to as *accessory minerals*.

While most of the silicates originate in igneous and metamorphic rocks, the clay minerals are formed within soils or are inherited from parent materials such as lacustrine deposits and shale.

Surface area of parent materials

The specific surface area of the constituent particles in the parent material determines the amount of interaction that is possible with the environment, particularly with water. Consolidated rocks have an extremely small surface area when compared with alluvial sands which in turn have a smaller area than clays, thus surface area increases as particle size decreases. Variations in surface area and particle size distribution have a profound effect on the speed of soil formation for it is found that soils will develop in sediments much more quickly than from consolidated rock of the same composition.

Permeability of parent materials

The permeability of the parent material influences the rate of moisture movement which in turn influences the speed of soil formation. The most permeable materials and therefore those that allow free movement of moisture usually have a high content of sand but as the particle size decreases the general tendency is for the material to become more impermeable so that clays allow only a very slow rate of moisture movement.

Perhaps compaction should be mentioned as a property of parent materials. Some superficial deposits are not usually regarded as consolidated but they may be compact and will restrict moisture movement. Thus a sandy deposit may have low permeability because of compaction. Another property that influences permeability is structure or the degree of organisation of the soil, as discussed on page 61. This is particularly important in connection with drainage and crop growth.

Climate

Climate is the principal factor governing the type and rate of soil formation as well as being the main agent determining the distribution of vegetation. The climate of a place is a description of the prevailing atmospheric conditions and for simplicity it is defined in terms of the averages of its components, the two most important being temperature and precipitation. Although averages are most commonly used, diurnal and annual patterns and extremes are not ignored since they give character and sometimes are important factors. For example, the occurrence of occasional high winds can determine the development policy of an area. However, it must be stressed that the atmospheric climatic data do not always give a true picture of the soil climate. For example, the amount of water in the soil may vary considerably within a distance of a few metres from permanently saturated to dry and quite freely draining, whereas there is virtually no difference between the amounts of precipitation at the two sites – one site may be in a depression where moisture can accumulate, and the other on an adjacent slightly elevated situation. Regularly these differences in the moisture regime at the two sites lead to the development of different soils and contrasting plant communities, with a dry habitat community on the elevated situation and a marsh community in the depression.

Temperature

Atmospheric and soil temperature variations are the most important manifestations of the solar energy reaching the surface of the earth, part of which is absorbed and converted into heat in the atmosphere and soil while the remainder is reflected back. The amount absorbed is influenced by the colour of the soil; since dark coloured soils absorb the most radiation they are the warmest. A proportion of the heat produced is maintained in the soil but some is lost to the atmosphere by convection of hot air from the soil and by back radiation. Also a considerable amount is used for evaporation of moisture into the atmosphere (Fig. 9).

Cloudiness, humidity, dust particles and pollution absorb radiation thereby reducing the amount reaching the earth's surface.

Vegetation has a buffering effect on soil temperature by absorbing and reflecting radiation during the day but during the night it reflects back to the soil some of the heat lost by radiation, so that temperature fluctuations are less beneath forests than in adjacent exposed sites. In a similar way a cover of snow reduces the loss of heat from the soil and may prevent the penetration of frost. On the other hand snow reflects over 90 per cent of the incoming radiation.

Perhaps it should be mentioned that only about 0·1 per cent of all energy reaching the earth's surface is absorbed by plants and fixed by photosynthesis and is therefore the total amount of energy that runs all the life on our planet.

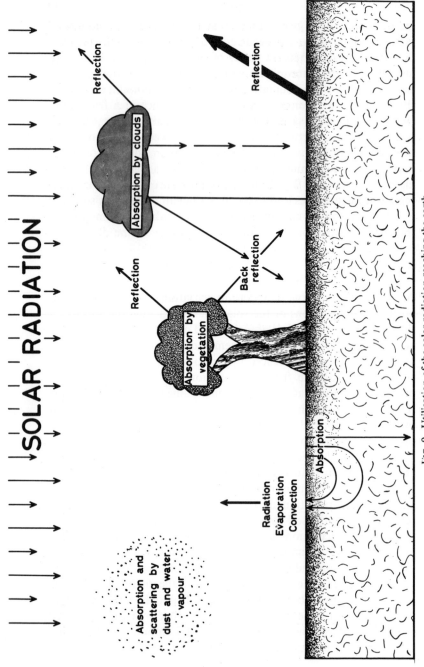

SOLAR RADIATION

Absorption and scattering by dust and water vapour

Reflection

Absorption by clouds

Reflection

Reflection

Absorption by vegetation

Back reflection

Radiation
Evaporation
Convection

Absorption

Fig. 9. Utilisation of the solar radiation reaching the earth

The main effect of temperature on soils is to influence the rate of soil formation, since for every 10°C rise in temperature the speed of a chemical reaction increases by a factor of two to three. The principal process in the soil to which this applies is the weathering of minerals. The rate of both biological activity within the soil and the breakdown of organic matter are also increased by a rise in temperature. In addition the amount of moisture evaporating from the soils is increased.

Development of vegetation can also be affected by soil temperature – in cool climates plants become active at 5°C and reach maximum activity at 20°C, the rate of maturation having increased three to four times.

The amount of radiation reaching the surface and soil temperatures are determined by a number of factors, probably the diurnal and seasonal variations are the most important for they give the soil well marked cycles. During the diurnal cycle in tropical and subtropical areas it is normal for heat to move downwards in the soil during the day from the surface, due to warming by incoming

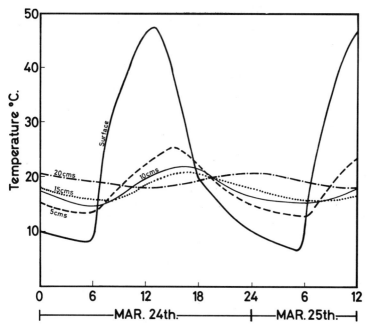

FIG. 10. Diurnal soil temperature variations at Cairo

radiation and upwards during the night as the surface cools. This takes place also during the summer period in the middle and higher latitudes, but in these areas during the winter the atmosphere is generally cooler than the soil and the incoming radiation is not sufficient to heat the soil which steadily cools and eventually may freeze from the surface downwards. The widest temperature fluctuations take place at the surface particularly in certain desert areas so that the surface shows both the absolute maximum and absolute minimum temperatures (Fig. 10).

Heat moves very slowly down through the soil so that the diurnal maximum in the lower horizons at about 20–30 cm occurs up to twelve hours after

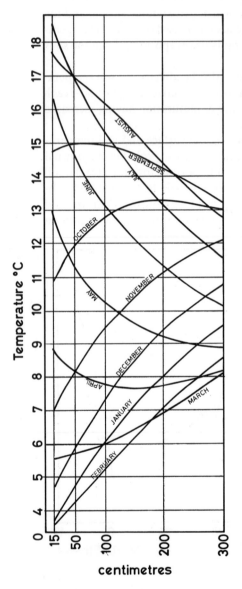

Fig. 11.

Annual soil temperature variations in southern England

the surface maximum (Fig. 10). This lag is greater in the annual cycle when the lower horizons attain their maximum even after the surface begins to cool in response to a seasonal change. The temperature fluctuations within the soil between seasons is greater at the surface than in the lower horizons. (Fig. 11). Thus during the summer in middle latitudes the diurnal mean surface temperature is higher than that in underlying layers but in winter the reverse is true.

Aspect and latitude influence soil temperature. Land surfaces normal to the rays of the sun are warmest but the greater the distance from the equator the smaller the amount of radiation reaching the earth's surface (Fig. 12).

With increasing altitude, the temperature decreases at the rate of about 1°C for each 170 m: on the other hand precipitation increases initially and then decreases. These two factors combine to produce a vertical zonation of climate, vegetation and soil as seen in all mountainous areas.

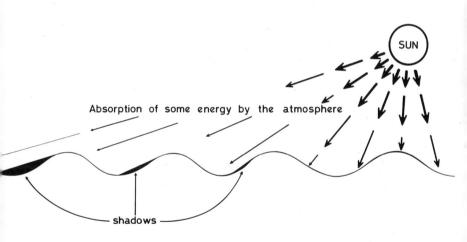

FIG. 12. The effect of aspect and latitude on the amount of solar radiation reaching the earth's surface. The intensity of the radiation is roughly indicated by the thickness of the arrows.

Moisture

The differentiation of horizons is determined very largely by the movement of moisture, therefore this process is of paramount importance. In fact, the soil solution might be regarded as the main "conveyor belt" in soils whereby ions and small particles are translocated from one place to another.

The moisture entering the soil is derived mainly from precipitation as rain and snow and contains appreciable amounts of dissolved CO_2. Thus it is probably more correct to think of moisture entering the soil as a dilute weak acid solution which is much more reactive than pure water. Some moisture takes part in a number of chemical reactions in the soil and some is retained, but by far the greatest amount is lost through drainage or by evapotranspiration, i.e., the

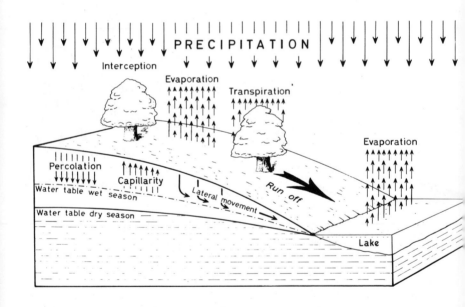

Fig. 13. The moisture cycle under humid conditions

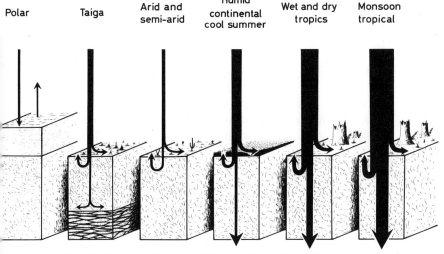

Polar | Taiga | Arid and semi-arid | Humid continental cool summer | Wet and dry tropics | Monsoon tropical

FIG. 14. The fate of moisture falling on the surface under a
number of different climatic conditions

combined processes of evaporation and transpiration. (Figs. 13
and 14.)

The intensity of precipitation varies from place to place over the
earth's surface and in areas of bare soil it is precipitation of moderate
intensity that is most effective in entering the soil. Light showers of
rain that hardly enter the soil are quickly lost by evaporation. Heavy
showers may cause moisture to accumulate at the surface particularly on clay soils leading to run off and creating an erosion
hazard. Precipitation can be intercepted by the foliage of vegetation
and is later lost by evaporation.

Generally moisture moves downwards after rainfall or melting
snow, but it can move upwards by capillarity in response to drying
out at the surface or it may move laterally through the soil on
slopes. Moisture movement within soils can also take place in
the vapour phase but this seems to be minimal (see page 77 *et seq.*).

One of the most fascinating aspects of climate is its continuous
change with time as revealed by an examination of the wide range
of geological strata. With regard to soils, the climate at any one
point on the surface of the earth has changed many times during
the last two million years leading to changes in the processes of

soil formation, so that most soils have been subjected to contrasting processes as determined by climatic changes (see pp. 31 and 124).

Organisms

Nearly every organism living on the surface of the earth or in the soil affects the development of soils in one way or another. The organisms can be considered under the following headings:

Higher plants	Microorganisms
Vertebrates	Mesofauna

Higher plants

Higher plants influence the soil in many ways. By extending their roots into the soil they act as binders and so prevent erosion from taking place, with grasses being particularly effective in this role. Roots can also grow within cracks in rocks forcing them apart. When plants die and their roots decay, they leave a network of passages through which water and air can circulate more freely.

One of the greatest contributions of the higher plants is through the addition of organic matter or litter to the surface. The total amount added by the different plant communities is very variable but it is no guide to the amount present in the soil which depends more upon the rate and type of breakdown. Tropical plant communities contribute as much as 25 tha^{-1} year^{-1}, tall grass prairie 5·0 tonnes and pine forest 2·5 tonnes but the amount of organic matter in the soils beneath these communities is in the reverse order. Pine forests may have an accumulation of about 15 cm of organic matter at the surface. Prairie grass soils have up to 15 per cent organic matter incorporated in the mineral soil while the soils of tropical rain-forests often contain less than 5 per cent organic matter.

Plants extract water and nutrients from the body of the soil and under natural conditions return most of the nutrients to the surface in their litter which decomposes and releases them, rendering them available for re-absorption (see page 120). The type of plant community can sometimes be used to assess soil conditions. Rushes usually indicate wet soils while heather grows best on dry acid soils.

Vertebrates

A few mammals including rabbits, moles and the prairie dog burrow deeply into the soil causing considerable mixing, often bringing subsoil to the surface. The classical examples of this are the crotovinas found in many chernozems, especially those of Europe where the blind mole rat is principally responsible (Fig. 15).

Uncontrolled grazing by animals such as goats will devour the vegetation and leave the surface bare for erosion. This is a conspicuous feature in many countries bordering the Mediterranean Sea and many semi-arid areas (Figs. 50 and 51).

Microorganisms

The predominant microorganisms are bacteria, actinomycetes, algae and fungi. The bacteria are the smallest and most numerous of the free-living microorganisms in the soil, where they number several million per gram with a live weight of 1000–6000 kg/ha in the top 15 cm. This weight is slightly less than that of the fungi but greater than that of the other microorganisms combined (Fig. 16).

The distribution of microorganisms in soils is determined largely by the presence of a food supply, therefore they occur in the greatest

FIG. 15. Mole hills in a field of cultivated grass. Each mole hill is about 30 cm in diameter and 15 cm high

numbers in the surface horizon which is a teeming mass of biological activity. Most members require an aerobic environment and have optimum temperature requirements of 25–30°C. It should be mentioned that these high temperatures occur in only a few soils, therefore microorganisms usually operate below their optimum.

Microorganisms are divided into two groups, the *heterotrophs* and the *autotrophs*. The former, including most of the bacteria, actinomycetes and fungi, obtain their food and energy from plant and animal remains, while the latter derive their body carbon solely from the carbon dioxide of the atmosphere. Therefore the heterotrophs are principally responsible for the decomposition of the litter.

The most important autotrophs are those that derive their energy from a variety of oxidation processes including the oxidation of ammonia to nitrites, nitrites to nitrates (see pages 42, 45) hydrogen sulphide to sulphur, sulphur to sulphate and ferrous iron to ferric.

Mesofauna

This group includes earthworms, nematodes, mites, springtails, millipedes, some gastropods and many insects, particularly ter-

FIG. 16. Thin section showing fungus decomposing organic matter

mites. Like microorganisms their distribution is determined almost entirely by their food supply and therefore they are concentrated in the top 2 to 5 cm; only a few, such as earthworms penetrate below 10 to 20 cm. Generally the mesofauna require an aerobic environment with conditions around neutrality but many can live in either acid or alkaline soils. The concentration of each organism varies greatly from place to place, but it is estimated that under optimum conditions the biomass of earthworms is about 80 g/m^2 and nematodes about 5–20 million/m^2.

The mesofauna are concerned largely with the ingestion and decomposition of organic matter, in addition many earthworms, termites and millipedes ingest both mineral and organic matter and as a consequence they produce faecal material which is a homogeneous blend of these two substances. They also transport material from one place to another and in so doing they produce passages which improve drainage and aeration. (Figs. 17, 18, 19.) Some

FIG. 17

A termitarium
in western Nigeria

mesofauna, particularly the nematodes, transmit virus diseases or are themselves parasites of great economic importance such as the potato eelworm.

FIG. 18. Soil profile with termites' nest near the bottom

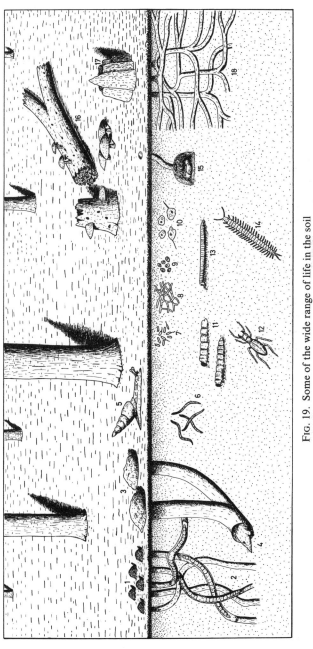

FIG. 19. Some of the wide range of life in the soil

1. Earthworm casts
2. Earthworms and their passages
3. Mole hills
4. Moles and their burrows
5. Snail
6. Nematodes
7. Bacteria
8. Fungal mycelium and actinomycetes
9. Algae
10. Protozoa
11. Insect and other larvae
12. Beetles and other arthropods
13. Millipedes
14. Centipedes
15. Lizard's eggs
16. Dead tree trunk with termite passages and fungal growths
17. Termitarium
18. Termite passages

Topography

This includes the dramatic mountain ranges and the flat featureless plains, both of which give the impression of considerable stability and seem to be timeless. However, this is not the case, for it is known from many investigations that all land surfaces, even those in areas of hard rocks such as granite, are constantly changing through weathering and erosion. Thus topography is not static but forms a dynamic system, the study of which is known as *geomorphology*.

Topography influences the soil in many ways. For example, the thickness of the soil is often determined by the nature of the relief (Fig. 20). On flat or gently sloping sites there is the tendency for material to remain in place and for the soil to be thick but as the slope increases so does the erosion hazard, resulting in thin stony soils on strongly sloping ground. Topography also influences the drainage and the amount of moisture in the soil.

Topographic features are produced by three main processes, tectonic processes (crustal disturbances), erosion and deposition. Initially all topographic features are produced by tectonic processes; the surfaces are then acted upon by running water, ice, frost and wind which are the principal agents of erosion and deposition.

The material removed by running water is normally transported to rivers to form alluvial deposits or taken to the sea to form deltas.

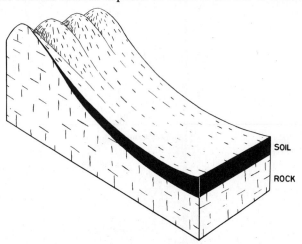

FIG. 20. Effect of topography on soil depth; with decreasing inclination the soil increases in thickness because the site is more stable

Erosion by ice in the form of glaciers is relatively local at present but was more widespread in the past when they produced characteristic erosional forms such as U-shaped valleys and depositional features such as drumlins and moraines that are common over large parts of Europe and North America (Fig. 21).

FIG. 21. Stratified glacial deposits in eastern Scotland. The two lower strata are unsorted compact till with angular stones and boulders. The upper stratum is a glacio-fluvial deposit hence the higher concentration of stones and boulders, many of which are rounded or sub-rounded

Wind action is confined largely to arid and semi-arid areas and to coastal positions where the formation of sand dunes is a most characteristic and conspicuous process. However, in the past wind activity was responsible for the silt deposits (loess) which are so common in parts of Australia, China, the U.S.S.R., the U.S.A. and elsewhere (Figs. 22, 23).

Repeated freezing and thawing of wet unconsolidated material on slopes causes a considerable amount of movement. This process is known as solifluction and is probably the principal geomorphological process taking place in polar areas and may lead to the accumulation of thick deposits of material on the lower parts of slopes and the formation of terraces.

The movement *en masse* of material down slopes is probably more common than is realised. This may be gradual and may therefore pass unnoticed but when it is rapid as in the form of landslides following earthquakes or torrential rain the amount of material that moves can be considerable and catastrophic (Fig. 24).

FIG. 22. Sand dunes in the Sahara

Fig. 23.

A loess deposit in central Poland. Note the characteristic prismatic structure in the upper part of the section and the dark coloured buried soil near to the base

Fig. 24. A landslide. The material is a laminated lacustrine clay which slid over a highway

Time

Soil formation is a very long and slow process requiring thousands and even millions of years. Since this is much greater than the life span of an individual human being, it is impossible to make categorical statements about the various stages in the development of soils.

Not all soils have been developing for the same length of time but most started their development at various points during the last hundred million years.

Some horizons develop before others, especially those at the surface which may take only a few decades to form in unconsolidated deposits. Horizons requiring a considerable amount of rock weathering may take more than a million years to develop. Three examples of soil formation or the development of soil through time are given on page 46 *et seq*.

Soil evolution

Initially pedologists tended to interpret most soil features as the result of the interaction of the prevailing environmental conditions at the time when soil examinations were made. However, it is now evident that most places have experienced a succession of different climates which have induced changes in the vegetation and in soil formation. Therefore, most soils are not developed by a single set of processes but undergo successive waves of pedogenesis. Furthermore each wave implants certain features that are inherited by the succeeding phase or phases. The occurrence of strongly weathered soils in the lower Saharan and Australian deserts are clear indications of some of the climatic and soil changes that have taken place, since these soils are considered to have formed initially under hot humid conditions but now occur in deserts. Similarly, many of the soils of Europe and North America show the influence of the cold conditions that existed when glaciers extended further to the south.

In historical times many evolutionary paths have been directed by man's activities; the polders of Holland and the puddled soils to grow rice in the Far East quickly spring to mind. Many plant communities are produced by the activity of man; excellent examples are the *Ericaceous* heaths of Europe and many of the savannas of the tropics, both of these plant communities being maintained by systematic burning.

Some workers, particularly Jenny (1941), have tried to demonstrate quite unconvincingly that these factors are independent variables *i.e.* each of them can change and vary from place to place without the influence of any of the others. Only time can be regarded as an independent variable; the other four depend to a greater or lesser extent upon each other, upon the soil itself or upon some other factor. For example, it is now generally accepted that the various plant communities are largely a function of climate which is a function of wind currents, latitude, proximity to water and elevation.

Many attempts have been made to show that some factors are more important than others and therefore play a major role in soil formation. Such efforts are a little unrealistic since each factor is absolutely essential and none can be considered as more important than any other, although locally one factor may exert a particularly strong influence.

3

Processes in the soil system

Soils are complex, dynamic systems, in which an almost countless number of processes are taking place. Generally these processes can be classified as chemical, physical or biological but there are no sharp divisions between these three groups. For example, oxidation and reduction are usually regarded as chemical processes but they can be accomplished by microorganisms; similarly the translocation of mineral particles can take place either in suspension or in the bodies of organisms such as earthworms.

Chemical processes

The main chemical processes include hydration, hydrolysis, solution, clay mineral formation, oxidation and reduction.

Hydration

Hydration is the process whereby substances absorb water. Few of the primary minerals undergo hydration directly, therefore very little takes place during the early stages of weathering. The principal exception is biotite which absorbs water between its layers, expands, and finally splits apart. Hydration is more often a secondary process affecting decomposition products such as iron and aluminium oxides.

Hydrolysis

Hydrolysis is probably the most important process participating in the destruction of minerals and soil formation. It is the replacement of cations such as calcium, sodium and potassium in the structure of the primary silicates by hydrogen ions from the soil solution eventually leading to the complete decomposition of the minerals. The products of hydrolysis such as calcium are then available to be taken up by plants or removed by water flowing through the soil or they may precipitate out of solution.

Solution

There are only a few substances found in soils that are soluble in water or carbonic acid. Nitrates, chlorides and sulphates are very soluble but these only occur in appreciable amounts in the soils of arid areas. Calcite and dolomite are less soluble but are widespread and form the major components of limestone, chalk and some other parent materials. These materials are very distinctive since they are almost completely soluble in carbonic acid and therefore supply only a very small residue after solution. Consequently soils developed on these materials are normally quite shallow. Less soluble is apatite (calcium phosphate) which can persist for thousands of years in some soils of humid areas developed in drift deposits. On the other hand, most other materials, particularly the silicate rocks, furnish a considerable residue of primary and secondary products.

Some minerals such as quartz which are usually considered to be inert and insoluble do dissolve eventually. This accounts for the small amount of primary material $<50\,\mu m$ found in many of the very old soils of humid tropical and subtropical areas.

Products of hydrolysis and solution

These include the weathering solution, the resistant residue and alteration compounds. The weathering solution contains the basic cations together with some iron, aluminium and silicate ions which are partly or completely lost, redistributed in the soil system or taken up by plants. The resistant residue includes quartz, zircon, rutile and magnetite which alter only very slowly but do decompose when present as very small particles.

The alteration compounds are principally hydroxides and oxides of iron and aluminium; silica and clay minerals. Iron forms ferric hydroxide, goethite – $\alpha FeO\text{-}OH$ or hematite – Fe_2O_3. Ferric hydroxide is an amorphous yellowish-brown substance that occurs in many soils in their initial stages of formation. Goethite is crystalline with reddish-brown colour but changes to yellowish-brown as it becomes hydrated. Goethite has a wide distribution, ranging from the tropics to the arctic and is one of the main colouring substances in soils. Hematite is bright red and occurs chiefly in soils of tropical and subtropical areas or in old geological formations. The other iron oxide is lepidocrocite which is bright orange in colour, occurring principally in soils subjected to periodic waterlogging.

In a crystalline form, aluminium oxide occurs mainly in the soils of humid tropical and subtropical areas as gibbsite $\gamma Al(OH_3)$ which is also the principal constituent of bauxite.

Amorphous or hydrous silica, in addition to forming part of the clay, may be lost in the drainage water or redistributed within the soil system. Sometimes it may accumulate and cement the soil into a massive rock-like material known in Australia as silcrete.

Finally there is manganese dioxide which is of restricted distribution being found as the blue black coating within the soil or associated with iron oxides and hydroxides in certain concretionary and massive deposits.

Transformation of individual minerals

The breakdown of the individual minerals depends largely upon the nature of the climate, thus the products of decomposition of a given mineral will vary from place to place. For example, feldspars can be changed to mica in a cool climate, to kaolinite in a hot, moderately humid climate and to gibbsite in a very hot, very humid climate. This wide variability in the end products of weathering applies also to the amphiboles and pyroxenes.

Clay minerals – their formation and properties

Early workers visualised a simple transformation from primary minerals such as feldspars to clay minerals like kaolinite but it has been shown that the structures of these two minerals are so different that this is not possible. For many situations it seems that the primary minerals undergo fairly complete decomposition to simple substances followed by the synthesis of the individual clay minerals. This involves principally the rearrangement of the silicon-oxygen tetrahedra and the aluminium octahedra both of which become aligned to form sheets within the clay minerals.

There are five main types of clay minerals important in soils; namely, *kaolinite, montmorillonite, hydrous mica, vermiculite* and *allophane*. The first four are crystalline and can be regarded as small sheet-like particles (Fig. 8). Allophane is amorphous or very finely crystalline with an indeterminate composition but generally has about equal amounts of aluminium hydroxide and silica.

One of the most important properties of clay minerals is that they have negative charges which allow them to adsorb and exchange cations on their surfaces i.e. they have a *cation exchange capacity* (see page 69), and because of variations in their structure they have widely differing cation exchange capacities. Also they have a capacity to absorb water. Kaolinite has a low cation exchange capacity (3–15 me/100g) and expands very slightly when wet. On the other hand, the exchange capacities of vermiculite (100–150 me/100g) and montmorillonite (80–150 me/100g) are high and both of these minerals can adsorb water and expand, particularly montmorillonite. Hydrous mica occupies an intermediate position with a moderate cation exchange capacity (10–14 me/100g) and a small capacity for swelling when wet. The occurrence of these minerals in soils is related largely to pH. Kaolinite forms in acid soils with mica, vermiculite and montmorillonite forming under progressively more alkaline conditions.

Flocculation and dispersion are two more properties of clays. Flocculation is the reaction whereby the individual particles of clay coagulate to form floccular aggregates. This can be demonstrated in the laboratory by adding a small amount of calcium hydroxide to a suspension of clay. Immediately floccular aggregates form and gradually settle to the bottom of the vessel – a similar reaction takes place in soils to form crumbs and granules. At the other extreme is the state of dispersion in which the individual particles are kept separate one from the other by many ions particularly sodium. Thus depending upon the nature of the cations present in the soil it may either be in a flocculated or dispersed and often massive condition (see page 62).

Clay minerals scarcely ever occur in a pure form in soils; often a single particle is composed of interstratified layers of two or more different clay minerals, or they may be enmeshed in large quantities of oxides – hence the many difficulties encountered in their identification.

Oxidation and reduction

It is convenient to consider these two processes together since one is the reverse of the other. Iron is the principal substance affected by these processes for it is one of the few elements that is usually in the reduced state in the primary minerals. Consequently when it is released by hydrolysis and enters an aerobic atmosphere it is quickly oxidised to the ferric state and precipitates as ferric hydroxide to give yellow or brown colours. If, on the other hand the iron is released into an anaerobic environment it stays in the ferrous state. Such soils range in colour from blue to grey to olive to black depending upon the precise compound that is formed. Vivianite (ferrous phosphate) imparts blue colours while black colours are due to sulphides which often form in coastal and estuarine marshes. The formation of horizons by partial or complete reduction is often referred to as gleying.

Physical processes

The main physical processes are translocation, aggregation, freezing and thawing and expansion and contraction, but the agencies responsible are very varied.

Aggregation

Aggregation is the process whereby a number of particles are held or bound together to form units of varying but characteristic shapes. This property is discussed on page 61.

Translocation

Many of the processes of soil formation and horizon differentiation are concerned primarily with removal, reorganisation and re-distribution of material in the upper 2 m or so of the earth's crust.

Percolating through the soil in a humid environment is a large volume of water which on moving downwards takes with it dissolved material some of which may be translocated to a horizon below or it may be lost in the drainage water. On slopes, some moisture moves laterally causing the soils at lower positions to be enriched by soluble substances. Sometimes, the whole soil may become saturated by water moving laterally and there is free water at the surface. Such situations are known as *flushes* and usually carry a plant community commonly indicative of a moist habitat.

Fine particles and colloidal materials are often transported in suspension from one place to another within the soil system. Perhaps the most important manifestation of this process is the removal of particles $<0.5\mu m$ from the upper horizons of some soils followed by their deposition in the middle position to form cutans (see page 53). Ultimately this can lead to differences of more than 20 per cent in the clay content between adjacent horizons.

As stated earlier many members of the mesofauna and some small mammals are responsible for redistributing large amounts of material within the soil.

Freezing and thawing

These two processes take place to varying degrees over a wide area of soils. During freezing a number of ice patterns develop as determined by a number of factors. Needle ice usually forms near to the surface of the ground and is responsible for heaving stones to the surface and the breakdown of large clods, hence the reason for ploughing before the onset of winter. Below the surface, freezing causes the formation of ice lenses and in polar areas repeated freezing and thawing causes profound disturbance of the soil and the development of a number of characteristic patterns such as stone polygons and mud polygons (Figs. 25 and 26)

Freezing and thawing also causes rocks to be shattered leading to the development of extensive areas of angular rock fragments and as already mentioned on page 28 a considerable amount of solifluction takes place in polar areas as a result.

Expansion and contraction

These processes take place mainly as a result of wetting and drying and are very important in soils containing a high proportion of clay with an expanding lattice such as montmorillonite. They occur principally in the soils of hot environments with alternating wet and dry seasons. Contraction is probably the more important process for it leads to the formation of wide and deep cracks causing the roots of the vegetation to be stretched and broken. During expansion high pressures are developed within these soils, causing rupture and the slippage of one block over the other and the formation of polished faces or *slickensides* on the blocks. Expansion and contraction also

Fig. 25.
Stone polygons 2–3 m wide, on Baffin Island

Fig. 26.

A mud polygon 120 cm wide from northern Alaska

cause the formation of micro-topograpnic features known as *gilgai*
(Fig. 27). They can also cause the disruption of the foundations of
buildings.

FIG. 27. Gilgai. The pressures produced by expansion and
contraction cause contortions in the soil and heaving of the
parent material to the surface

Biological processes

The most important biological processes taking place in soils are the
decomposition of organic matter or humification, nitrogen trans-
formations and the translocation of material from one place to
another.

Humification

The decomposition of organic matter or humification is an extremely complex process involving various organisms including fungi,
bacteria, actinomycetes, worms and termites. It has been shown
that in the case of pine needles the process is very slow with invasion

and decomposition by the fungus *Lophodermium pinastri* starting when the needles are still on the tree and continuing after they have fallen onto the ground. This fungus causes the formation on the needles of characteristic black spots and transverse black bars which are clearly seen in hand specimens (Fig. 28). When the needles are on the ground they are attacked by wave after wave of fungi, each wave being fairly specific in its action and each successive wave being capable of breaking down more complex plant compounds than the previous wave (Fig. 16). At first the simple compounds such as starches and sugars are decomposed, followed by the proteins, cellulose, hemicellulose and finally the very resistant compounds such as tannins. Gradually the material is decomposed and the dark coloured substance known as *humus* is formed, the whole process taking about seven to ten years.

Small arthropods such as mites also play a part in the process by eating the softer inner parts of the needles as well as consuming some of the fungal mycelium. For these reasons the organic matter beneath pine forests shows progressive decomposition with depth. In contrast the decomposition of the leaves of deciduous trees by earthworms and bacteria is usually much more rapid and may be accomplished within a year.

FIG. 28.

Pine needles infected with the fungus *Lophodermium pinastri* as indicated by the characteristic black spots and bars

The true character of humus is still uncertain but it is now generally believed to be a product of microbial synthesis and plant decomposition and is composed of heterogeneous polymers formed by the interaction of polyphenols, amino acids and other substances.

The decomposition of organic matter is also accomplished by other organisms including earthworms and termites. These are much more vigorous than fungi and bacteria for they ingest and decompose the litter soon after it has fallen, therefore soils containing these organisms usually have little organic matter on their surfaces.

Generally the formation, decomposition and redistribution of organic matter can be represented by the carbon cycle in Fig. 29.

Nitrogen transformations

The main forms of nitrogen transformation are *ammonification*, *nitrification* and *nitrogen fixation*. Ammonification is the process whereby nitrogenous compounds in plant and animal tissues are decomposed to produce ammonia which is changed by nitrification into nitrite, and then into nitrate, each stage being accomplished by specific microorganisms. The formation of ammonia is accomplished by heterotrophic bacteria but the two other stages are brought about by autotrophic bacteria. Ammonia is oxidised by *Nitrobacter*, *Nitrosomonas* and *Nitrococcus* and the nitrite is oxidised by *Nitrobacter*. These processes require aerobic conditions; if the soil is waterlogged for any length of time the nitrogenous compounds are reduced by denitrification to nitrogen which is lost to the atmosphere.

Nitrogen fixation is the process during which atmospheric nitrogen is utilized by soil bacteria to form their body protein. The organisms include *Azotobacter, Clostridium pasteurianium* and *Beijerinckia* which upon death enter the nitrogen cycle and are decomposed to form nitrate for plant uptake.

There are also a number of bacteria which enter the roots of certain plants, particularly members of the *Leguminoseae*. There they multiply, form nodules and fix atmospheric nitrogen which then passes into the conducting system of the plant as an essential element.

Since nitrogen does not form part of the lithosphere all the nitrogen that forms part of the tissues of plants and animals must

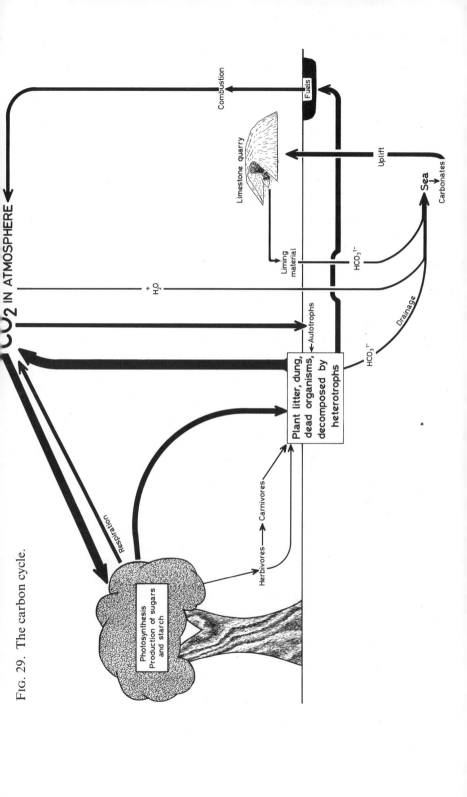

Fig. 29. The carbon cycle.

come initially from the atmosphere, being transformed in the soil by various microorganisms. Therefore, in one sense, nitrogen fixing microorganisms can be regarded as the corner stone of **all life.**

The above processes may appear to be of little significance in the direct formation of soils but they are of paramount importance in plant nutrition. Each of the above processes can be regarded as part of a cycle during which nitrogen ions pass from one stage to another (Fig. 30).

Translocation

Some emphasis must be given to those biological processes that bring about churning and translocation. The most dramatic manifestation of this process is brought about by soil-inhabiting vertebrates but probably the greatest amount is accomplished by earthworms and termites. (Figs. 15, 17, 18, 19).

In conclusion it must be stressed that soil-inhabiting fauna and flora contribute enormously to the biological balance in nature for without them the remains of plants and animals would accumulate to very great thicknesses. This is clearly demonstrated by the formation of peat which is the accumulation of organic matter under wet anaerobic conditions which are not suitable for litter decomposing organisms; hence its accumulation (Fig. 64).

Among the organisms in the soil there exists an extremely complex inter-relationship for seldom does a single type of organism exist or operate separately from the others. Some highly contrasting organisms coexist while others are predators, competitors or parasites. Earthworms and bacteria coexist, for when earthworms ingest a mixture of organic and mineral material they also take in large numbers of microorganisms, particularly bacteria, because the alimentary systems of earthworms do not seem to have a distinctive bacterial flora but is the same as that of the surrounding soil. These bacteria, together with the enzymes liberated in the alimentary system of the worm are responsible for the breakdown of the organic matter, thus supplying energy and body tissue for the worm. Eventually the worms die and are themselves decomposed by other microorganisms or they may be eaten by a mole or other organism.

Another example is for plant tissues to be softened by fungal growth and then eaten by small arthropods which at the same time may eat some of the fungus. Later when the arthropod dies it will be

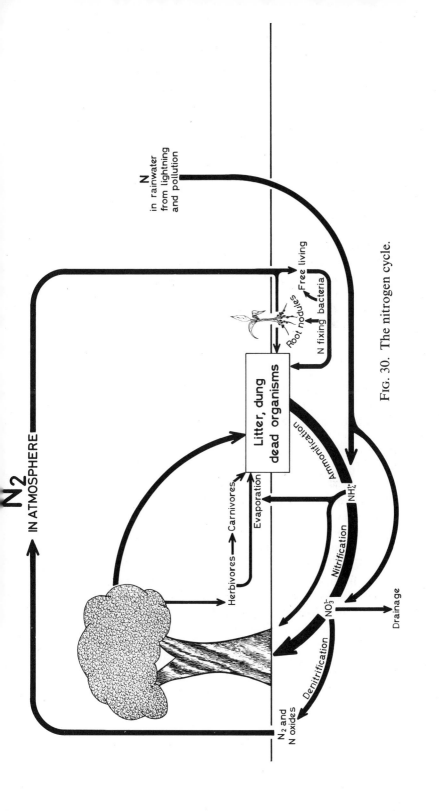

Fig. 30. The nitrogen cycle.

decomposed by other microorganisms. Some nematodes live on protoplasm from living cells but they can be trapped and killed by a special loop mechanism of certain fungi and then decomposed by the fungus to form its food.

There are also the protozoa which seem in part to be scavengers by devouring small fragments of organic matter, but they also consume large amounts of bacteria and may control their numbers.

A fascinating example of the indirect effect of microorganisms is found in Western Australia. There, the soils and weathered rocks are many tens of metres thick and over a period of many millennia they have accumulated soluble salts, particularly sodium chloride. The salts are not harmful when the native Jarrah forest remains undisturbed but large areas are being attacked and destroyed by the root invading fungus *Phytophthora cinnamomi*. This allows more moisture to enter and percolate through the soil thereby leaching the salts from the high to the low ground and into streams and rivers. Thus the forest is being destroyed, the soils in the lower parts of the landscape are becoming saline and unsuitable for most plants, and domestic water supplies rendered useless. This devastation should be contrasted with the nitrogen fixing bacteria which are the cornerstone of life. (Removal of the forests by man has the same effect.)

Thus it is seen that there are a number of different but interrelated facets to life in the soil.

Soil formation

The above factors and processes combine to form the very wide range of soils that occurs on the surface of the earth, and is discussed in Chapter 7. The general trends are summarised and presented diagrammatically in Figs. 30 and 31. In addition the formation of a podzol from a sand dune, a krasnozem from rock and a chernozem from loess will be considered. These three specific examples illustrate some of the more important principles in soil formation. Podzols form in cool humid conditions in which translocation and accumulation of sesquioxides in the middle horizon is the main process. In the krasnozems progressive weathering under hot humid conditions predominates, while chernozems form in cool semi-arid areas where there is vigorous faunal activity and translocation of carbonates.

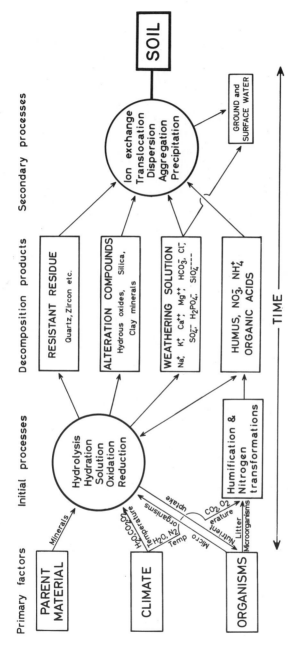

Fig. 31. Soil formation (adapted from Yaalon 1960)

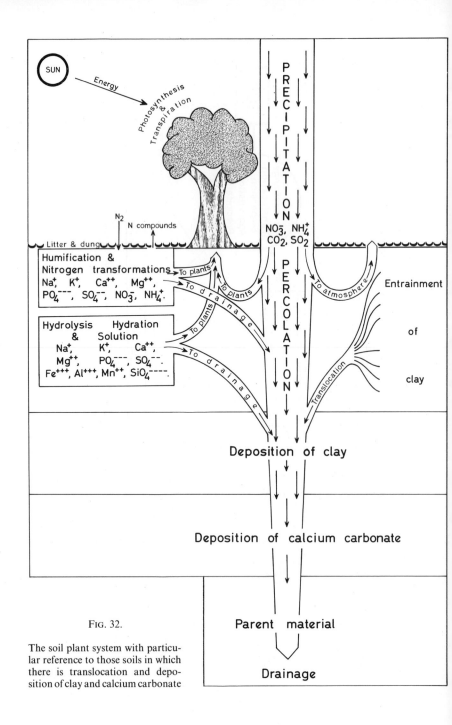

Fig. 32.

The soil plant system with particular reference to those soils in which there is translocation and deposition of clay and calcium carbonate

Stages in the formation of a podzol – Fig. 33

Initially there is the primosol stage in which there is a thin litter at
the surface overlying a crude mixture of organic and mineral material
about 10–15 cm thick. Below this is the unaltered sand. This stage

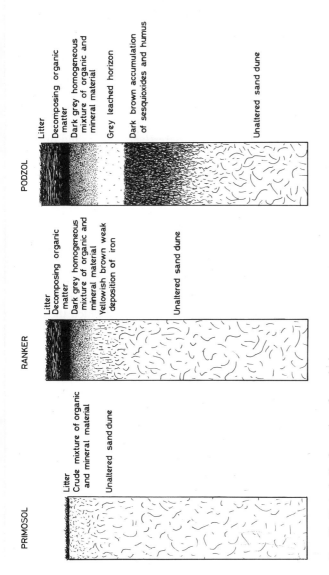

FIG. 33. Stages in the formation of a podzol

is of short duration and is quickly followed by the accumulation of a thick litter with progressively more decomposed organic matter with depth. Below, there is a dark grey homogeneous mixture of organic and mineral material up to 15cm thick and followed by some rusty staining of the sand. This staining is due to the deposition of iron and marks the initial stage in the formation of the middle horizon. This is then a ranker with only a well developed upper horizon. The next stage is the fully developed podzol taking less than a thousand years to form and characterised by the development of a marked bleached horizon due to the leaching out of iron, aluminium and humus followed by their deposition below. In some cases sufficient material accumulates to cause cementation and hardening of the horizon. These three stages may take place at the same site or they may be found in progressively older sand dunes. Accompanying the development of the soil there is usually a plant succession starting with mosses, lichens, grasses and herbs and culminating in pine forest.

Stages in the formation of a krasnozem – Fig. 34

At stage 1 there is the fresh rock surface followed by the formation of rankers in small depressions where material can accumulate. Gradually the rock weathers during which some of the products remain while others are lost in solution from the system, hence the lowering of the surface. This is the altosol stage at which the soils are still fairly shallow and contain a high content of primary minerals. With time the soil deepens, and all of the primary minerals such as feldspars, amphiboles and pyroxenes are decomposed and the iron is oxidised to give the characteristic red colour of the krasnozem at stage 4. In contrast to podzols, krasnozems develop very slowly because of the large amount of hydrolysis involved and may take over one hundred thousand years to form. There is also a plant succession on these soils with tropical rain forest as the final stage.

Stages in the formation of a chernozem – Fig. 35

Probably the greatest areas of chernozems are developed in loess which is a fine wind-blown sediment deposited during the last glaciation. Following the deposition of loess there is colonisation of

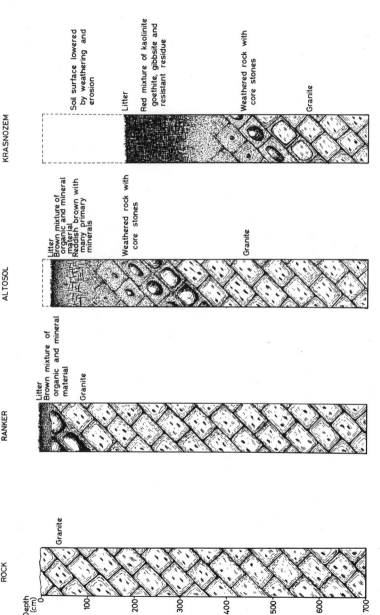

FIG. 34. Stages in the formation of a krasnozem

the site by vegetation and the formation of a thin crude mixture of organic matter and mineral soil to give a primosol. This is followed by a denser growth of grass and the formation of a ranker in which there is a fairly homogeneous blend of organic and mineral material resulting from a vigorous earthworm and blind mole rat population. There is also a small amount of translocation of calcium carbonate. Ultimately a deep chernozem forms mainly through the continued activity of the soil fauna incorporating organic matter and mild leaching causing the translocation of calcium carbonate. Like podzols, chernozems will form within a thousand years.

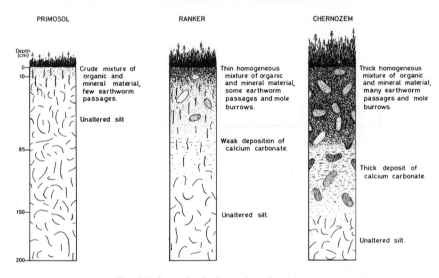

FIG. 35. Stages in the formation of a chernozem

4

Properties of soils

Because soils develop as a result of the interaction of the factors of soil formation they have a wide range of properties. The most common are listed and discussed below. Methods for describing and estimating some of the properties are given in Appendix II.

1. Thin section morphology
2. Colour
3. Particle size distribution
   Large separates >2mm
   The fine earth <2mm
4. Handling properties
5. Texture
6. Structure and porosity
7. Atmosphere
8. Moisture status
9. Density
10. pH
11. Organic matter
12. Cation exchange properties
13. Soluble salts
14. Carbonates
15. Elemental composition
16. Amorphous oxides
17. Concretions

1. Thin section morphology

Resulting from the pioneering work of Kubiëna (1953) thin sections of soils are being used in an increasing number of investigations, and it seems probable that in the future this will form one of the major techniques in soil investigations.

Thin sections of soils are prepared by a technique similar to that employed by geologists for making thin sections of rocks, but soils must be impregnated with a resin and hardened before they can be cut, ground and polished.

With this technique, investigations can be conducted on small details of the type of material and degree of organisation or fabric of the soil.

Among the many features seen in thin sections is the redistribution pattern of the clay fraction to form thin layers or *cutans* which coat the aggregates in some soils (Fig. 36). Further, the structure of soil can be seen and studied better in thin sections than in hand

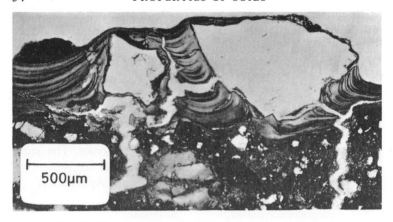

FIG. 36. Clay cutans bridging pore space and surrounding sand
grains

FIG. 37.

Thin section of granular structure in the upper horizon of a
vertisol. There are abundant subrounded peds

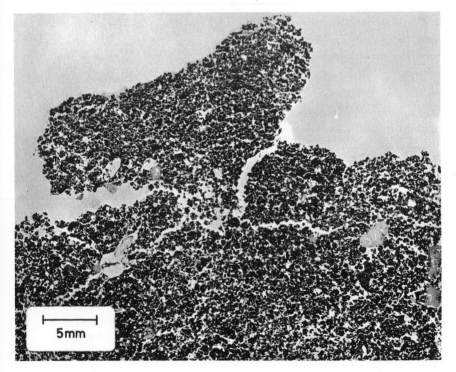

specimens in the field (Figs. 37 and 38). Weathering and other mineralogical investigations can also be carried out and it has a high potential for microbiological investigations, allowing observations to be made on organisms in their natural habitats (Fig. 16).

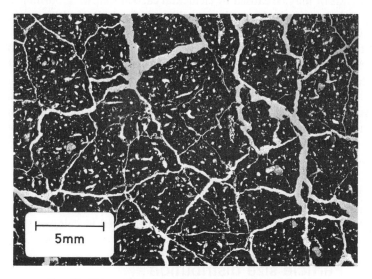

FIG. 38. Thin section of angular blocky structure with a high content of ovoid and circular pores within the individual peds

2. Colour

A very high proportion of the names of soils is based upon colour, since this is the most conspicuous property and sometimes the only one that is easily remembered. Further, many inferences made about soils are based upon colour.

Generally the colour of a soil is determined by the amount and state of iron and/or organic matter. Hematite is responsible for the red colour of some soils of tropical and subtropical areas. However, the mineral responsible for most of the inorganic colouration of freely drained soils is goethite which has colours that range from reddish brown to yellow as its degree of hydration increases. The highly hydrated yellow and yellowish-brown forms are sometimes referred to as limonite.

Grey, olive and blue colours occur in the soils of wet situations and originate through the presence of iron in the reduced or ferrous state. The colour of the upper horizons usually changes from brown to dark brown to black as the organic matter content increases. Dark colours are produced also through the presence of manganese dioxide or may be caused by elemental carbon following burning.

Pale grey and white originate through the lack of alteration of light-coloured parent materials, deposition of calcium carbonate, efflorescence of salts or the removal of iron leaving significant amounts of uncoated light coloured minerals such as quartz, feldspars and kaolinite.

Some horizons have a colour pattern which may be mottled, streaked, spotted, variegated or tongued. Possibly the most common and important colour pattern is yellow and brown mottles on a grey background which is interpreted as resulting from seasonal wetting and drying of the horizon.

Although the colours of most horizons are produced by pedogenic processes there are a number of instances when they are inherited from the parent material; for example, many sediments of Devonian and Permo-Triassic age are bright red in colour.

3. Particle size distribution

Soil particles are divided initially into two size classes with the limit normally set at 2 mm to delimit the "fine earth" from the larger separates including gravel.

Larger separates – material > 2 mm

A knowledge of the nature and properties of the coarse particles can often lead to important conclusions about the origin and formation of the parent material and about the soil itself. In some superficial deposits the sand and silt fractions have a similar mineralogical composition to that in the material > 2 mm. Therefore the larger separates can be used to assess the chemical composition of some soils and parent materials. This is a common field practice in many glaciated areas.

The shape of the stones often gives a clear indication of the processes which have influenced the formation of the parent material and/or the soil itself. Rounded stones occur in alluvium and beach deposits, subangular to subrounded stones in glacial drift, angular stones result from exfoliation or frost action.

Stones in superficial deposits are orientated in a number of specific patterns. The stones in alluvium usually have an imbricate pattern with their long axes aligned in the direction of river flow and dipping upstream. Glacial deposits often have stones aligned in the directions of ice flow. In areas of vigorous frost action the stones become orientated by frost heaving to form a number of patterns at the surface as well as in the soil itself (Fig. 25). On a flat site the stones within the soil are vertically orientated whereas on a sloping situation they become oriented parallel to the slope and normal to the contour. A similar type of orientation on slopes is associated with certain forms of soil creep found in many tropical areas.

When wind is an agent of erosion and deposition, stones at the surface become faceted by the fine particles blown against them. Such stones are known as *ventifacts*.

The fine earth – material <2 mm

The material <2 mm is divided into sand, silt and clay, the size limits of which vary between workers but normally either the International scheme or that proposed by the United States Department of Agriculture is adopted. The former has four fractions whereas the latter is more detailed and has seven fractions as set out below in Table I.

TABLE I: *Size limits of soil separates*

U.S. Department of Agriculture scheme		International scheme	
Name of separate	Diameter (range)	Separate	Diameter (range)
	Millimetres		Millimetres
Very coarse sand	2·0 – 1·0		
Coarse sand ..	1·0 – 0·5	Coarse sand.. ..	2·0 – 0·2
Medium sand ..	0·5 – 0·25		
Fine sand	0·25– 0·10	Fine sand	0·20– 0·02
Very fine sand ..	0·10– 0·05		
Silt	0·05– 0·002	Silt	0·02– 0·002
Clay	<0·002	Clay	<0·002

According to the relative amounts of sand, silt and clay twelve classes have been created and presented in the form of a triangular diagram by using only three size limits as shown in Fig. 39.

The mechanical composition of soil horizons is the result of fairly specific processes. Soils developed in fairly recent deposits have inherited a considerable proportion of their characteristics from the parent materials but as soils increase in age, more and more clay is formed and they gradually become increasingly fine. However, the translocation of clay particles from one horizon to another may cause the horizon losing the clay to become coarser and the receiving horizon to become finer.

The shape of the particles in the sand fraction is often a useful guide to the origin of the material. Sand varies in shape from smooth and round to very rough and angular. The former is found in wind blown material and beach sand whereas rough angular sand occurs in glacial deposits. The fine earth contains both primary and secondary minerals. The former usually dominate the sand fraction

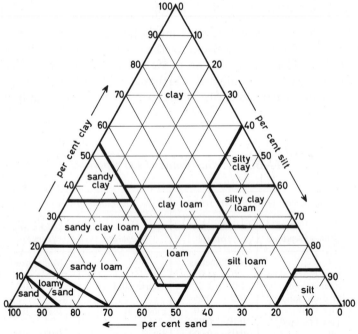

FIG. 39. Triangular diagram relating particle size distribution to texture according to the U.S.D.A.

and it has been found that the mineralogy of the fine sand fraction gives a fairly accurate picture of the nutrient reserve in the soil.

The clay fraction is composed predominantly of crystalline clay minerals and amorphous material. The study of this fraction yields important data, allowing important conclusions to be made about the processes taking place in the soil. The accurate determination of the particle size distribution in soils is a fairly long and tedious process but a simple approximate estimate can be carried out as given on page 147.

4. Handling properties or consistence

When soils are manipulated between the fingers and thumb they exert varying degrees of resistance to disruption and deformation as determined by their mechanical composition, degree of aggregation, content of organic matter and moisture content (see page 147). Generally the pressure needed to disrupt a dry soil increases with the content of fine material, sands are usually quite loose while clays form very hard aggregates. Moist sands have a small measure of coherence whereas moist clays are plastic and become very sticky when wet, particularly if the content of montmorillonite is high. The presence of large amounts of humified organic matter in the soil is particularly important for it increases the plasticity of sandy soils but has the reverse effect on a clay by reducing the stickiness. The consistence of soils of medium texture does not change very much with variations in moisture content. In either the dry or moist state they are usually friable, i.e. firm with well formed aggregates that crumble easily when pressure is exerted on them. When wet, these soils tend to be slightly sticky but never to the same extent as clays.

Some horizons that are massive and hard offer a considerable degree of resistance to disruption as a result of cementation by substances such as iron oxides, aluminium oxide and calcium carbonate. Resistance to disruption can result also from physical compaction.

The consistency of a soil is a very important agricultural property and it is essential for the soil to have the correct consistence at the time of cultivation. If it is too dry and hard undue strain will be placed on the implements, on the other hand if it is too wet and sticky the implements may stick and the soil may become puddled thus producing a poor seed bed for crops.

5. Texture

The texture of the soil refers to the "feel" of the moist soil resulting from the mixture of the constituent mineral particles and organic matter. Therefore it is an approximate measure of the particle size distribution or mechanical composition which is measured in the laboratory. Whereas particle size data are required for certain studies, texture as determined by feel is often more closely related to the behaviour of the soil in the field and to the physical properties of significance in agriculture.

Texture is determined by rubbing the moist soil between the fingers and thumb and is commonly determined in the field. It is a subjective technique but can be mastered with some experience and being a simple technique it has an advantage over particle size determination, which is a long tedious process.

Although different soils may have the same texture they may not have the same particle size distribution and *vice versa*. This is due mainly to variations in the amounts of organic matter, type of clay, shape of particles and degree of aggregation. It is usual for the upper horizons to contain varying amounts of organic matter. When the content is small the effect is minimal but large amounts of organic matter cause the soil to be smooth and to appear to have a higher silt content.

The type of clay will also have an effect since some clays (montmorillonite) absorb more water than others (kaolinite); thus two soils may contain the same amount of clay but the one with montmorillonite will be more sticky and plastic than that with kaolinite.

The shape of the particles can also be important. For example, when the sand grains are round the grittiness will appear to be less than if they are angular.

In some soils the clay particles are cemented to form small aggregates which cause the soil to appear to have a higher silt content. Thus a soil which has the particle size of a clay has the feel of a silty clay.

The assessment of texture and textural classes is given on page 147.

6. Structure and porosity

This refers to the degree and type of aggregation and the nature and distribution of pores and pore space. In many soils the individual particles exist as discrete entities but in others the most common arrangement is for the particles to be grouped into aggregates with fairly distinctive shapes and sizes. These aggregates are known as *peds*. The range is from those very sandy soils in which each particle is separate through those with well formed aggregates to those that are compact and massive. In soils with well formed aggregates or when there is mainly sand the individual units may have only a few points of contact but generally are surrounded by a continuous pore phase. On the other hand in massive soils it is the mineral material that forms the continuum but it may and often does contain discrete pores.

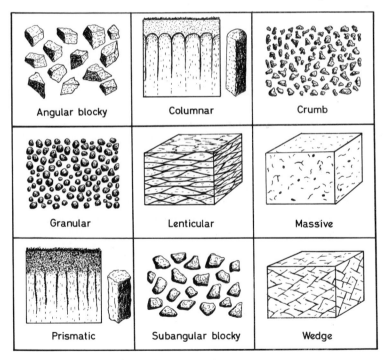

FIG. 40. Some types of soil structure

Up until recently the description of structure has been based largely on hand specimens. With the development of a technique for preparing large thin sections of soils a fuller appreciation of structure is possible (Figs. 37 and 38).

The principal types of structures recognised in the field are described below and illustrated in Fig. 40:

Angular blocky: Peds of irregular shape that have sharp angular corners, flat, convex and/or concave faces. They are usually firm or hard and fairly tightly packed, common in many middle horizons.

Columnar: Vertically elongated peds with domed upper surfaces and three or more flat vertical faces. The peds usually grade into the underlying material and may be composed of smaller peds. This is the characteristic structure of the middle horizon of certain alkaline soils (Fig. 41).

Crumb: The peds have irregular shapes and rough surfaces forming a loose porous mass, common in upper horizons beneath grassland (Fig. 42).

Granular: Subspherical peds usually forming a fairly loose mass; common in upper horizons.

Lenticular: Lens-shaped peds with convex surfaces; they overlap each other and are common in many horizons that have been frozen or compacted by implements.

Massive: Continuous soil phase without peds or continuous pore space, usually found in lower horizons.

Prismatic: Vertically elongated peds with three or more vertical flat faces. The peds often have somewhat indeterminate upper and lower boundaries. This structure is common in many medium and fine textured horizons (Fig. 43).

Single grain: Occurs in very sandy horizons so that the individual grains are separate one from the other and do not form aggregates.

Spongy: A tangled sponge-like mass of organic material.

Subangular blocky: Peds of irregular shape with convex and/or concave faces and rounded corners and fairly tightly packed; common in many middle horizons.

Vermicular: A maze of intertwining faunal passages which are filled or partially filled with vermicular aggregates that are usually faecal in origin; common in upper horizons with vigorous earthworm or termite activity.

Wedge: Wedge shaped peds formed by the intersection of planar pores at 30°–60°. Usually occurs in middle horizons containing a large amount of montmorillonite and formed by expansion and contraction.

Structure is one of the least permanent properties of soil for it

FIG. 41. Columnar structure – the width of the section is 1 m

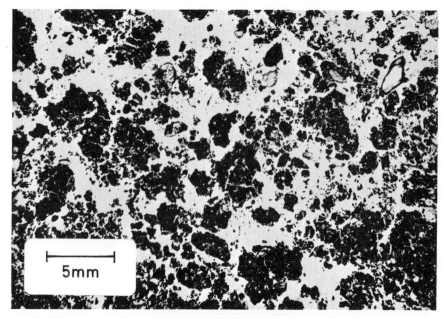

Fig. 42. Thin section of crumb structure composed of
numerous loose porous aggregates

Fig. 43.

Prismatic structure in
the middle horizon.
Note the sharp edges
and flat faces of the
prisms

can be altered very rapidly by cultivation or any other type of disturbance. It is also a very important property, thus at present a considerable amount of research is being devoted to the formation and retention of structures such as crumb or granular. Soils with a granular or crumb structure allow free percolation of excess moisture and at the same time roots are free to grow in the pore space between the peds. Soils with massive or prismatic structure severely restrict root development because there is little pore space into which roots can grow and often they are anaerobic.

7. Atmosphere

The soil atmosphere occupies the pores and pore spaces in soils and is generally continuous with the above ground atmosphere but differs from it in many ways. The soil atmosphere usually has a much higher content of water vapour and a much higher content of carbon dioxide which may vary from 0·25 per cent to 5 per cent as compared with 0·03 per cent in the above ground atmosphere. The atmosphere in the soil is constantly circulating and interchanging with that above ground. If it did not circulate the content of carbon dioxide and other gases would increase to proportions toxic to plant roots.

8. Moisture status

The overall annual moisture status of soils is inferred from morphological characteristics and on limited numbers of measurements made at various times of the year. On this basis the following five classes of drainage for the whole soil are usually recognised. The forms in which water occurs in soils are given on page 88.

Excessively drained

Water moves rapidly through the soils which have bright colours due to oxidising conditions.

Freely drained (Colour Plates IIIA & IVB)

Water moves steadily and completely through the soil with little tendency to be waterlogged. Such soils also have bright colours due to oxidising conditions.

Imperfectly drained

The soil is moist for part of the year with one or two horizons showing mottling owing to extended periods of wetness and reduction of iron to the ferrous state. The wetness in the soil as in the other two classes below may be caused by an impermeable horizon, high water-table or high precipitation.

Poorly drained

The soil is wet for long periods of the year with the result that many of the horizons are mottled, with at least one that is blue or grey resulting from reducing conditions.

Very poorly drained (Colour Plate IIA)

The soil is saturated with water for the greater part of the year so that most of the horizons are blue or grey caused by reducing conditions. Peat may also be present as a consequence of the high degree of wetness.

9. Density

The true density of a soil is a measure of the density of the constituent components and varies from about 2·65 for the mineral particles to about 0·2 for the organic matter. On the other hand the density of the whole undisturbed soil or *bulk density* is

$$\frac{\text{Wt. of a dry block of soil}}{\text{Vol. of block when sampled}} = \frac{g}{cm^3} = x$$

This takes into account the density of the soil materials themselves and their arrangement or structure. Therefore a loose porous soil will have a smaller bulk density than a compact soil even although the density of the individual particles in the two soils may be the same.

Whereas the bulk density of most cultivated soils is about 1·3 the extremes can vary from 0·55 for soils developed on volcanic ash to 2·0 for some strongly compacted lower horizons. This property is attaining some importance in fertility studies because continuous cultivation by heavy implements increases the bulk density by inducing compaction which reduces percolation and root pene- tration.

10. pH

The pH of a solution is a measure of its acidity or alkalinity and is defined as the negative logarithm of the hydrogen ion concentration. Soils do not behave like simple solutions therefore it is not possible to give an accurate definition of soil pH but for many purposes it can be considered as similar to the above. The range for soils is normally from pH 3 to 9. Very low values are found in the soils of marshes and swamps that contain pyrite or elemental sulphur. At the other extreme very high values result from the presence of sodium car- bonate. Within the normal range the two principal controlling factors are organic matter and the type and amount of cations. Large amounts of organic matter induce acidity except when counterbalanced by high concentrations of basic cations. Acidity is also induced by large amounts of aluminium in solution. Generally pH values about neutrality are associated with large amounts of exchangeable calcium and some magnesium, sometimes supple- mented by free carbonates. As discussed later, pH is a very important factor in plant growth.

11. Organic matter

Up to the present the exact nature of the organic matter in soils has not been determined. Therefore in routine analysis of soils only the total amounts of carbon and nitrogen are determined.

The carbon percentage is usually multiplied by the conversion factor of 1·72 to give an indication of the total amount of organic matter present. The ratio of carbon to nitrogen (C/N ratio) is calculated and used as a measure of humification and ranges from > 100 to 8. The former high figure is for fresh litter and peat while the latter occurs in many upper horizons containing a mixture of mineral material and well humified organic matter.

A rough measure of the amount of organic matter in soils can be attained by igniting dry soil and determining the loss in weight (Appendix II). This technique is applicable to most soils except those containing carbonates which decompose when ignited.

The total amount of organic matter in soils varies from < 1 per cent to > 90 per cent. The latter figure is for the relatively unaltered material that occurs in the litter at the surface and in peat. Normally the upper horizons contain < 15 per cent organic matter and over very large areas they contain < 2 per cent, even when the supply of litter to the surface is high as in certain humid tropical areas.

The decomposed organic matter which is usually well mixed with the mineral material in the upper horizons is the *humus* and has unique properties which very largely determine the character of the upper horizons. Firstly, humus is capable of absorbing large quantities of water thus increasing the water holding capacity of the soil and is therefore of importance in crop production. Like clays, humus has a Cation Exchange Capacity (C.E.C.), but it is considerably higher, being about 300 me/100 g and therefore increases considerably the cation holding capacity of the soil. Humus can be dispersed or flocculated depending upon the nature of the cations present and as already stated on page 59 it influences handling consistence. Thus humus behaves somewhat like clay but it is easily destroyed by microorganisms hence the difficulties encountered in keeping it at a high enough level in many soils. In addition it has already been stated that organic matter affects soil colour and it supplies essential elements when it is decomposed.

Under natural conditions the humus content of a virgin soil is usually higher than in adjacent cultivated areas. This is caused by a higher rate of addition of organic matter by the natural vegetation accompanied by a lower rate of biological activity and lower temperatures.

12. Cation exchange properties

The two most important cation exchange properties are the cation exchange capacity and the percentage base saturation.

The C.E.C. of the whole soil is a measure of the exchange capacity of the clay and humus expressed as milligram equivalents per 100 grams of soil i.e. me/100 g. The range is from about < 5 me/100 g for some lower horizons up to > 100 me/100 g for upper horizons containing high percentages of organic matter, vermiculite or montmorillonite.

The percentage base saturation is a measure of the extent to which the exchange complex is saturated with basic cations. The general trend is for the amount of exchangeable base to increase with decreasing rainfall and for calcium to be dominant but sodium may be dominant in certain arid regions. Conversely low figures indicate intense leaching but changes can be induced very easily by cultivation. Continuous cultivation leads to a rapid reduction in cations such as calcium and potassium if they are not added in fertilizers.

When the total content of exchangeable cations are expressed as me/100 g the amount may seem small but when recalculated on the basis of the amounts of cations available to a growing crop there are usually several thousand kg of cations per hectare.

13. Soluble salts

When a soil is shaken with water and filtered, the filtrate will contain some dissolved salts but they only occur in significant proportions in the soils of arid and semi-arid areas (see page 117). There they accumulate because the annual precipitation is insufficient to leach the soils or because the water-table is at a shallow depth and moisture is drawn to the surface by capillarity bringing with it dissolved salts which are left behind as the moisture evaporates (Fig. 44). Flooding by sea water also causes salinity in soil but this is of minor importance except in countries such as Holland which depend upon large areas reclaimed from the sea. The predominant anions are bicarbonate, carbonate, sulphate and chloride while the cations include sodium, calcium, magnesium and small amounts of potassium.

These ions occur in widely varying proportions and depending upon the particular ratio they impart a number of properties to the soil, some of which are detrimental to plant growth. Soils can be grouped into two scales depending upon their salinity and alkalinity.

Degree of salinity	% by weight
Free of salts	$<0.15\%$
Slightly saline	$0.15–0.35\%$
Moderately saline	$0.35–0.65\%$
Strongly saline	$>0.65\%$

Fig. 44. Salt efflorescence on the surface of the soil and a partly exposed solonchak profile

Degree of alkalinity	% of total exchangeable ions
Slightly alkaline	<20% exchangeable sodium
Moderately alkaline	20–50% exchangeable sodium
Strongly alkaline	>50% exchangeable sodium

A common feature of many soils containing soluble salts is that microelements accumulate to toxic proportions. In the case of boron concentrations over $1 \cdot 0 \, \mu g/g$ are toxic. Generally plants cannot tolerate a salt concentration of over $0 \cdot 5$ per cent and as the concentration rises much above this limit the percentage of halophytes such as *Atriplex* spp and *Beta* spp increases rapidly.

There are immense areas of potentially very fertile land that have a high concentration of salts but in many cases the salts can be removed, rendering the land suitable for agriculture as discussed on page 117.

14. Carbonates

Carbonates of calcium and magnesium, particularly the former, are widely distributed in soils, occurring separately or they may be associated with other salts. The most important properties of carbonates are: (1) They are relatively easily soluble in water containing dissolved carbon dioxide, and therefore can be quickly lost or redistributed within the soil. (2) When present in an amount as small as 1 per cent of the soil they can dominate the course of soil development because this amount is sufficient to raise the pH value over neutrality and sustain a high level of biological activity. (3) Carbonates, particularly calcium carbonate, are the first substances to start accumulating as the climate becomes arid. (4) Both calcium and magnesium are essential plant nutrients. (5) Carbonates are regularly added to many arable soils to raise their pH values for optimum plant growth.

15. Elemental composition

The principal elements occurring in soils and their proportions in the earth's crust have already been given (page 11) and only in very detailed studies is it necessary to determine the total amount of each element present. Where soil formation has proceeded for a long period and where there has been a considerable amount of hydrolysis and solution, it is usual to perform a quantitative analysis to estimate the ten or so dominant elements in each horizon as well as in the underlying parent materials. This is to determine the predominant chemical changes that have taken place during soil formation. The elements which are determined include silicon, aluminium, iron, sodium, potassium, calcium, magnesium, manganese, titanium, zirconium, nitrogen and phosphorus. However, the determination of the total amount of certain single elements is performed very frequently, particularly nitrogen and phosphorus because of their importance as essential plant nutrients.

16. Amorphous oxides – free silica, alumina and iron oxides

The type, amount and distribution of amorphous materials can be used as criteria for measuring the degree and type of soil formation. Also they are regarded as being formed during the current phase of pedogenesis and, therefore, are valuable criteria for differentiating between relic and contemporary phenomena. When performing the determinations it is customary to treat the soil with an extracting reagent such as sodium dithionite, citric acid or sodium pyrophosphate and to determine the amount of each constituent in the extract. Published data pertaining to these techniques can be challenged but certain trends have been discovered which seem to be reasonably valid. Some of these trends are given in Chapter 6.

Although chemical techniques have been used almost exclusively for the study of these oxides, other methods, including the study of soils in thin sections by the electron microscope and electron probe, are yielding many new and interesting data. For example, it has been shown in some instances that silica often forms small aggregates while alumina and iron oxides form coatings on clay minerals and other surfaces.

17. Concretions

Under certain conditions some constituents form local concentrations which may become very hard. For example, iron may accumulate to form concretions in certain soils of the tropics and some periodically waterlogged soils. Similarly, calcium carbonate and gypsum form concretions or massive horizons in arid and semiarid areas.

5

Soil fertility and land use

What is a fertile soil? This is a difficult question to answer since the requirements of plants vary considerably. For example, their moisture requirements differ widely. Plants, such as barley, maize or sugar cane need well aerated soils for optimum growth. On the other hand, some rice plants require wet anaerobic conditions for the early part of their life cycle.

Soil fertility is usually discussed in the context of crop production but it can be considered from the point of view of inherent soil fertility and induced soil fertility. Nearly every soil has a certain inherent fertility. Soils that are wet, acid, alkaline or deficient in a particular element will support a specific plant community. Therefore they can be regarded as fertile with regard to the plants growing on them, but when man wants to replace one of these natural plant communities with a crop, the inherent fertility may not be suited to the particular crop. Then it may be necessary to change the soil to induce the type of fertility to suit the needs of the crop. Nevertheless, it is possible to set out in general terms the plant-soil requirements and to state the factors affecting these requirements.

Factors affecting plant growth:
1. Root-room and root-hold
2. Aeration
3. Moisture
4. Temperature
5. Essential elements
6. pH
7. Stable site

1. Root-room and root-hold

Most plants extend their roots into the soil for anchorage as well as to extract nutrients and water. Although the greatest amounts of roots occur within the top 15 cm of the soil some can penetrate to depths of over 4 metres. Therefore the thickness of soil available for root penetration is very important. Soils may be shallow because of rock or a compacted horizon near to the surface or due to water-logging caused by a high water-table. Whereas it is usually impossible to increase the thickness of a soil over rock, it is often possible to lower the water-table by drainage, using the techniques mentioned below. Thus drainage increases the volume of soil available to the plants. However, there is a limit to such an increase due to difficulties in removing the excess water (Fig. 45).

With crops such as barley, maize and sugar cane, root-room restricted by rock usually means a reduction in the potential nutrient supply and available moisture. On the other hand with tree crops there is an additional factor of poor stability due to shallow rooting, thus trees grown on such soils are very liable to be blown over.

Drainage

The drainage of a soil can be achieved by a number of relatively simple techniques. In most temperate countries mole drainage or tile drainage is used. Mole drainage is carried out by drawing through the soil at a given depth a bullet shaped object the "mole" which forms a continuous passage into which water can percolate and move. The passages are laid out to connect with a main ditch which carries the water away. Tile drainage involves digging a trench and laying short lengths of clay pipe end to end and then filling the trench. This is a more permanent and effective method than mole drainage and is the main method used in Britain. In a number of cases mole drainage is used to supplement tile drainage. At present, tests are being conducted to evaluate continuous lengths of perforated plastic pipes which are laid by a machine that draws the pipe into the soil without having to dig a trench. If the tests are successful a considerable saving in the cost of drainage could result.

In countries with very heavy rainfall and much run-off it is customary to use open ditches. These are extremely effective, but they need a considerable amount of maintenance. Further, they can induce erosion if they are not well laid out.

In some cases toxins are produced under anaerobic conditions. Ethylene is produced in wet soils and inhibits root growth while cyanide is formed in the roots of peach trees causing their death (Greenwood 1970).

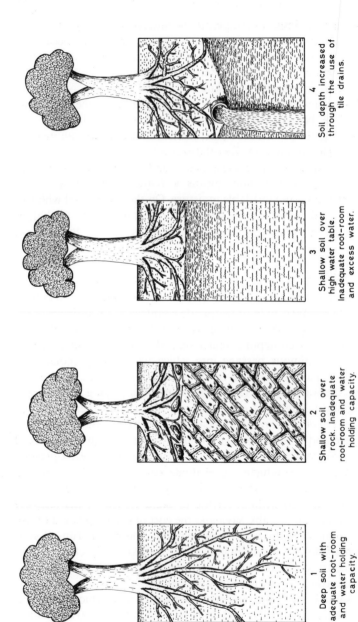

1

Deep soil with
adequate root-room
and water holding
capacity.

2

Shallow soil over
rock. Inadequate
root-room and water
holding capacity.

3

Shallow soil over
high water table.
Inadequate root-room
and excess water.

4

Soil depth increased
through the use of
tile drains.

FIG. 45. Some factors influencing the variations in the depth of soil

Drainage has a number of beneficial effects on the soil, it:
1. Increases the temperature in cool countries
2. Improves the structure
3. Allows air to enter
4. Increases the rate of organic matter decomposition
5. Improves the bacterial population
6. Improves germination
7. Allows a greater variety of crops to be grown
8. Reduces the incidence of plant disease such as blight
9. Causes fertilisers to be more effective and not washed out of the soil
10. Improves the health of livestock by reducing the incidence of foot rot and the frequency of the snail that is the intermediate host of liver fluke.

Perhaps it should be pointed out that while drainage is often beneficial to agriculture it may be at times an ecological hazard by threatening rare plant communities. For example, the drainage of dune slacks will threaten species such as *Baldellia ranunculoides* and *Epipactis palustris*.

2. Aeration

Plant roots and soil organisms require a constant supply of oxygen for respiration. Those plants such as rice that normally root in anaerobic conditions have a special mechanism for transferring oxygen from the stem and leaves to the roots. Since most roots and organisms develop in aerobic conditions they derive their oxygen from the soil atmosphere which usually contains an adequate supply (see page 65). A restricted or reduced supply of oxygen can be due to soil compaction, a high water-table or to the accumulation of carbon dioxide due to a lack of interchange between the soil and above ground atmosphere. Good aeration is usually facilitated by a granular or crumb structure and free drainage.

3. Moisture

An adequate and balanced supply of moisture is essential for plant growth. Moisture is constantly being taken up by plants together with nutrients and is lost by transpiration. It is estimated that 1 kg of dry weight increase in plants requires about 500 kg of transpired water, therefore a steady supply is necessary if growing plants are to remain alive. Under certain extreme conditions plants may lose more water than they take up even though there may be an adequate supply in the soil. This condition is known as physiological drought and occurs commonly during the day in very hot climates but the plants recover during the cool of the night.

The moisture in soils can be considered in terms of input, retention and losses.

Moisture input

The moisture entering the soil is derived from three main sources, rainfall, melting snow and irrigation. In humid climates the input by rainfall or from melting snow is usually adequate, but in arid and semi-arid areas an adequate system of agriculture can be sustained

Fig. 46. Furrow irrigation prior to cropping

only by irrigation which may take many forms. The water may be run on to a flat land surface causing complete flooding for a given period and then the excess water drained away (Fig. 46). Alternatively the water is run into furrows if the plant roots are likely to suffer from oxygen deficiency as a result of complete flooding and waterlogging (Fig. 46).

Sprinkler or overhead irrigation is often practised on small areas or when the ground is sloping and therefore unsuited to flooding. This system is usually portable and can be relatively inexpensive reducing the need to install and maintain ditches to conduct the water (Fig. 47). But these systems are wasteful of water because of high evaporation. Tube irrigation to individual plants is now being tried and has many advantages; water wastage is reduced to a minimum and in addition fertilizers can be added to the water as required.

Fig. 47. Sprinkle irrigation

Moisture retention

The moisture retained in the soil will depend upon the amount removed and the speed of removal. Consider a soil that is saturated with water. At first, water flows freely downwards through the pore spaces in response to gravity and is known as *gravitational water*. The amount of gravitational water is very important because it washes out some of the essential elements as well as translocating fine particles and is one of the main factors in horizon differentiation. When the excess water has drained away the soil is said to be at *field capacity* which is the maximum water retaining capacity. At this stage some of the moisture retained exists as thin films and bridges between particles and is known as *capillary water*, so called because it is capable of slow movement over surfaces in any direction and is partly available to plants (see Moisture Losses below). Finally, there is the physically and chemically bound water which is largely fixed and known as *hygroscopic water*.

Water will percolate rapidly through the soil if it is very porous, through being very sandy or because of a well developed structure, thus the retention is likely to be very low. Fine textured and organic soils have smaller pore spaces and the particles themselves can absorb moisture, therefore moisture retention is higher and moisture movement is slower.

Thus, texture, organic matter content and structure affect the movement and retention of moisture in soils. Generally clays and organic soils have the highest moisture retaining capacity, while silts and organic soils have the highest available moisture. Clays retain more water than silts but a higher proportion is strongly held and therefore unavailable to plants.

Moisture losses

The moisture retained in the soil is lost mainly by evapotranspiration. Therefore the rate of loss will depend upon temperature and plant cover, so that as temperature and plant cover increase, moisture losses will also increase. However, only part of the capillary water retained in the soil is available to be taken up by plants which will wilt and die after the available moisture has been exhausted. This is known as the *wilting point*.

The water retaining capacity and wilting point of the soil can be measured by simple experiments (Appendix IV). It has been found that as the water retaining capacity varies so does the wilting point. Clay soils contain much more water at the wilting point than sandy soils.

It is probably correct to say that on the world scale, water is the main limiting factor to plant growth, for even in humid areas such as the east of England, supplementary irrigation in most years can accomplish a substantial increase in crop production. In many semi-arid areas where irrigation is not possible various methods of moisture conservation have to be practised such as dry-farming (see Glossary page 162).

4. Temperature

Most plants have their optimum growth within a specific temperature range. In cool climates germination will not start until the temperature is about 5°C. This restriction becomes increasingly important towards the poles with their long winters and cool spring weather.

On the other hand, high soil temperatures may result in an excessive loss of moisture with drought as a consequence.

5. Essential elements

There are a number of elements essential for plant growth. These can be divided roughly into the macroelements and the microelements. The macroelements are required in relatively large amounts whereas the microelements are required in only small amounts. Furthermore, these elements must be present in the correct proportions for when there is a deficiency or excess of any one element plant growth can be affected seriously and the plants develop symptoms of nutrient starvation or toxicity. Since animals and man live directly or indirectly on plants, sometimes they can develop deficiency or toxicity as a result of plants being affected.

The 16 elements essential for plant growth are:

Macroelements	Microelements
Carbon	Manganese
Hydrogen	Copper
Oxygen	Zinc
Nitrogen	Molybdenum
Phosphorus	Boron
Potassium	Chlorine
Calcium	Iron
Magnesium	
Sulphur	

Most of the elements are derived initially from the weathering of minerals and taken up by the plant roots. They are then returned to the soil in the plant litter which decomposes to release the elements which are again taken up by the plants. This cyclic process is fundamental in certain natural plant communities which are sustained purely by this process, excellent examples being some rain-forests.

The functions of most of the elements are not fully known but they affect plants in the following manner:

1. They form plant tissue
2. They act as catalysts and intermediates in a wide range of metabolic processes.

Each element plays a specific role as follows:

Carbon, hydrogen and **oxygen** are the major constituents of plant tissue and are derived from the atmosphere and water.

Nitrogen is derived from the atmosphere or dead tissues and in both cases it is transformed by bacteria in the soil into ammonia and nitrate which are taken up by the plant roots (see page 42). It occurs in great quantities in young plants particularly the leaves. Nitrogen forms a part of every living cell, occurring in chlorophyll and all proteins with many of the latter serving as enzymes. Abundance of nitrogen leads to green succulent growth while nitrogen deficiency causes a loss of colour, reduction in protein production and a gradual yellowing and stunted growth.

Phosphorus is a constituent of every living cell and occurs in the protoplasm, with its greatest concentration in seeds thereby increasing their production. Phosphorus deficiency causes a purplish colouration at the seedling stage with later yellowing, stunted growth and delayed maturity.

Potassium is essential in all cell metabolic processes although the exact nature is not known. Apparently, it influences the uptake of other elements and affects both respiration and transpiration. It also encourages the synthesis and translocation of carbohydrates thereby encouraging cell wall thickening and stem strength. A deficiency can cause lodging in cereals and yellowing of the leaf tips and margins.

Calcium in the form of calcium pectate forms part of the cell wall structure and is necessary for the growth of the meristem (see Glossary, page 165). A deficiency leads to malformation of the growing parts but symptoms are seldom seen in the field.

Magnesium is active in enzyme systems and forms part of the chlorophyll. A deficiency causes discolouration and sometimes, premature defoliation.

Sulphur occurs in some amino acids and also in the oils of some plants such as cabbages and turnips. A deficiency leads to stunting and yellowing.

Iron and **manganese** play a role in enzyme systems and are necessary for the synthesis of chlorophyll. The activity of these two elements is interrelated since iron can be inactivated by an excess of manganese. Iron deficiency only becomes evident in older leaves and is seen as a yellowing particularly in the intervein areas. This is known as *iron-chlorosis* and is seen most commonly on calcareous or alkaline soils.

Manganese deficiency is similar to that of iron but the chlorosis is usually more marked with the whole of the intervein area losing its green colour.

Boron appears to play a role in calcium utilisation and the development of the actively growing parts of the plant. A deficiency leads to such conditions as heart-rot of beet and internal cork of apples. It is also essential for the fixation of nitrogen by bacteria in the nodules of legumes.

Copper and **zinc** form part of the enzyme systems and are necessary for the formation of growth promoting substances. Copper deficiency is common on peat leading to growth abnormalities such as rapid wilting and weak stalks. Zinc deficiency symptoms vary from plant to plant; mottled leaf of citrus is a well known example.

Molybdenum is necessary for the reduction of nitrate in the plant, otherwise nitrate will accumulate and interfere with protein synthesis. Nitrogen fixation by legumes is also dependent upon molybdenum.

Chlorine is responsible for regulating the osmotic pressure and cation balance in plants.

Availability of essential elements

Soils can be thought of as the reservoir of most of the essential elements for plant growth. Soils that have high content of primary minerals such as feldspars, amphiboles and pyroxenes contain a large reservoir of elements but strongly weathered soils may have only a small reserve. Although the total reserve of elements is a factor in determining soil fertility, probably more important is the degree of availability of the elements since in a number of cases the elements such as calcium may be tightly locked up within the silicate structure or in undecomposed organic matter and be unavailable or only slowly available to plants. Therefore a clear distinction has to be made between the total and available elements.

Elements such as calcium and magnesium when present as carbonates are usually easily available because carbonates are readily soluble in the soil solution. Generally, the available cations occur as exchangeable cations on the surface of the clay particles from which they enter into solution or are taken up directly by plant roots. Thus, the higher the base saturation the more cations there are available. The anions essential for plants also vary very much in availability. Nitrogen usually occurs as a part of the tissue of organisms and is unavailable until it is transformed by microorganisms into ammonia and nitrate which are easily available. Phosphates which occur principally as calcium phosphates are very slowly soluble and thus not readily available. Therefore, in order to know whether a soil is fertile or not it is necessary to know the degree of availability of the essential elements.

Various laboratory methods have been devised for assessing the type and amount of available elements in the soil. In these methods the soil is usually treated with a weak acid such as acetic acid and the elements determined in the filtrate. Then, based upon the experience of the chemist it is possible to state which elements are deficient and also their degree of deficiency. This technique is very rapid, allowing several hundred samples to be assessed in a single day but the interpretation of the results is very subjective because it depends very largely upon the experience of the operators.

The best methods for assessing soil fertility are field experiments which are designed to determine which elements are deficient and also to assess the amount of each deficient element which has to be added to the soil. This method also takes into consideration the effects of variations in the environmental factors such as climate. For example, it may be necessary to determine the amount of phosphorus required by wheat. Then a number of plots are laid out and a crop is planted in each plot. This is followed by applying an adequate amount of the essential elements *except* phosphorus which is added at three or four different levels starting with no phosphorus and increasing in stages up to excess. After the crop has grown it is harvested and the grain production weighed. Then a graph is prepared by plotting the weight of grain against phosphorus added and from the graph it is possible to determine the amount of phosphorus required for optimum grain production. This method is more reliable, but it is time consuming, particularly with crops other than annuals.

When investigating the nutrient requirements of plant species it may not be possible to carry out field experiments so that the soil scientist and ecologist usually have to rely on pot experiments. This method of investigation is advantageous when it is necessary to exercise rigorous control over the environment and particularly when trials are being conducted on microelements. The technique involves the collection of soil samples which are dried, ground and sieved so as to remove the coarse particles and to make them homogeneous. The soil is then placed in pots and seeds planted. Varying amounts of nutrients can be added before or after planting depending upon the nature of the experiment. The plants are then usually grown in a controlled environment in a greenhouse.

Many pot experiments are conducted using pure quartz sand or some other pure medium, particularly when testing the effect of microelements, and therefore offer advantages over field trials. In addition there is a considerable saving in space allowing numerous experiments to be conducted at the same time.

A further point is that the nutritional requirements of plants vary widely from species to species. For example, root crops take about twice as much nitrogen, potassium and calcium as cereals thus the assessment of the availability of nutrients always has to be made with reference to a specific crop.

When crop cultivation is carried out, the reservoir of elements in the soil is either insufficient or their production by weathering and microbial processes is too slow and therefore it is usual to add fertilizers, compost or manures. These vary considerably in type, the precise one being determined to a certain extent by their availability and the nature of the crop. Set out below is a list of the common fertilizers and the elements that they supply.

Sulphate of ammonia	20·5% N
Urea	45% N
Rock phosphate	11–15% P
Super phosphate	7–8% P
Basic slag	2–8% P
Bone meal	7–13% P
Potassium chloride	39–42% K
Potassium nitrate	14% N; 37% K

Some fertilizers are more soluble in the soil solution than others, therefore some elements are more available than others, so that in a number of cases the total content of an element in a fertilizer may not represent the amount available. This is particularly true of the phosphorus bearing fertilizers which are slowly soluble.

Microelement deficiencies and toxicities

It is not usual to add microelements to soils because either they are released in sufficient amounts by the weathering of minerals or they may occur in sufficient amounts in fertilizers as impurities, but there are a number of well known situations with microelement deficiencies. For example, most plants growing on peat soils are deficient in copper because of its slow release by plant decomposition.

When discussing the deficiency and toxicity symptoms of elements it is not sufficient to consider them only in relation to plants. It is necessary, also, to consider them in the wider context of animals and man who depend directly or indirectly upon plants for food, since deficiencies or toxicities in plants can be passed on to animals and man. Many gardens near tanneries have been manured for long periods with composted leather waste which contains a high content of copper and chromium. These elements have been taken up in large amounts by the plants and passed on to the people eating the vegetables. It is suggested that the higher incidence of stomach cancer in such areas is due to the excessive amounts of these elements.

Toxicities can arise in a number of other ways. A particularly good example is the smoke from aluminium factories containing a high content of fluoride which may be toxic to the surrounding vegetation. Another example is the abnormally high content of lead in hedgerow soils due to the various lead compounds in petrol engine exhaust gases.

Some elements such as cobalt and selenium required by animals and man may not be necessary for plant growth. Nevertheless, it is essential for plants to take up these elements so that they can be passed on to the animals and man when they eat the plants (Fig. 48).

A

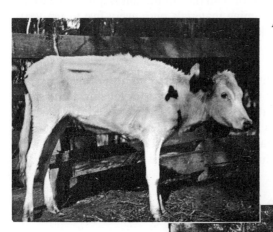

B

Fig. 48.

A. A calf with advanced symptoms of cobalt deficiency

B. The same animal on a balanced diet

Cobalt deficiency in cattle is common in many parts of the world, particularly good examples being found in Australia. In Scotland selenium deficiency causes muscular dystrophy in sheep.

6. pH

Whereas the pH values of soil are fairly variable (see page 67) the pH values of most cultivated soils ranges from 5·5–7·5 but for each crop there is a fairly narrow range. Crops such as tea grow best at pH values about 4·5, others such as wheat have an optimum range of 6–7·5. The pH of the soil is maintained at a suitable level by the addition of liming material which is usually a form of ground limestone.

Since the application of materials to the soil is quite expensive it is customary to add liming materials once every five to ten years. This is made possible by the fact that the liming material is lost relatively slowly by leaching from the soil because of its low solubility.

The degree of acidity of the soil affects a number of soil properties and processes, more especially the activity of the micro and meso-organisms. Those that are particularly beneficial to crop production, such as earthworms and bacteria, prefer conditions about neutrality. In some cases acid soils encourage the growth of certain plant pathogens such as club root which can be controlled by liming.

7. Stable site

In order to conduct a system of continuous crop production it is essential to have a stable site since continuous removal of material from the surface means the loss of the most fertile part of the soil as well as a steady and gradual reduction in soil thickness and root room. All sites are liable to erosion either by wind or water.

Wind erosion

Wind can remove particles of sand, silt and clay from any site and there are a number of very well known examples of catastrophic wind erosion such as the famous dust storms of the early 1930's in

Fig. 49. Dust storm

the midwestern prairie states of the U.S.A. This erosion was a result of removing the natural grassy vegetation and leaving the soil surface exposed for long periods between crops (Fig. 49). A common form of wind erosion is moving sand dunes which are most frequent in coastal situations. These are a hazard because they can be blown onto adjacent agricultural land or across roads. Sometimes the volume of material that moves is sufficient to bury buildings and it is claimed that a complete village was buried at Culbin in north-east Scotland. More recently wind erosion in southern England is partly due to the removal of hedges.

Water erosion

It is on slopes that erosion by water is the greatest hazard and usually sites of more than a few degrees require some erosion control measures. There are three main types of water erosion, as determined by the volume and the speed of the water moving over the surface. Firstly there is *sheet erosion* (Fig. 50) which takes place when water moves evenly over the surface at a fairly slow speed. This form of erosion is insidious and can go on for many years without being noticed by anyone except the expert. As the speed of movement increases the water begins to cut into the surface of the soil causing the second type – *rill erosion*. The third type is *gully erosion* which is caused by a large volume of water moving rapidly over the surface, particularly on moderate or steep slopes (Figs. 51 and 52).

FIG. 50. Sheet erosion. The surface of the soil has been washed away exposing the roots which may eventually die due to desiccation

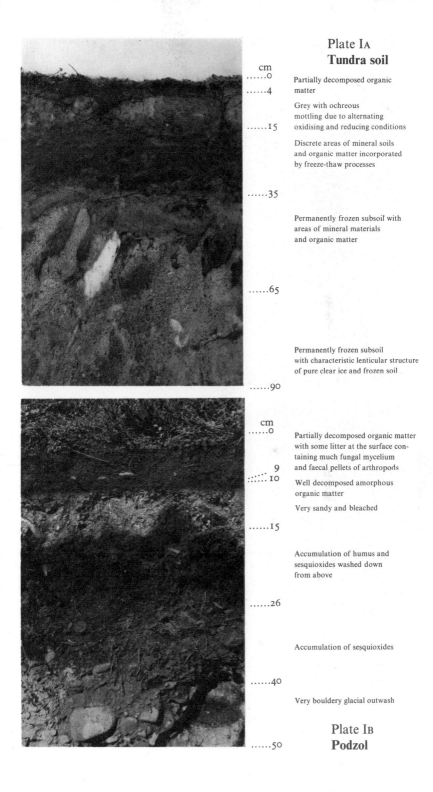

Plate IA
Tundra soil

cm
......0

......4

Partially decomposed organic matter

Grey with ochreous mottling due to alternating oxidising and reducing conditions

......15

Discrete areas of mineral soils and organic matter incorporated by freeze-thaw processes

......35

Permanently frozen subsoil with areas of mineral materials and organic matter

......65

Permanently frozen subsoil with characteristic lenticular structure of pure clear ice and frozen soil

......90

cm
......0

Partially decomposed organic matter with some litter at the surface containing much fungal mycelium and faecal pellets of arthropods

9
......10

Well decomposed amorphous organic matter

Very sandy and bleached

......15

Accumulation of humus and sesquioxides washed down from above

......26

Accumulation of sesquioxides

......40

Very bouldery glacial outwash

......50

Plate IB
Podzol

cm
......0

Plate IIA
Subgleysol

Very plastic, partially
decomposed organic
matter

......30

Grey with ochreous
mottling due to alter-
nating oxidising and
reducing conditions

......55

Blue grey and permanently
saturated with water

......90

cm
......0

Very plastic, partially
decomposed organic
matter

......15

Olive with faint ochreous
mottling due to wetness
caused by the impermeable
pan at 50 cm

......50 Thin iron pan

Solifluction deposit
derived from
weathered granite;
indurated

Plate IIB
Thin iron pan soil

......100

Fig. 51. Gully erosion. This is the normal type of devastation
that can result following the removal of the vegetation

Fig. 52. A deep gully that has undermined the tree leaving the
roots dangling in the air

The material removed by sheet erosion often accumulates on the lower parts of slopes so that it is common to find the top soil increasing in thickness down the slope. The material removed by rill and gully erosion often enters streams and rivers where it can cause much damage and create many problems. For example, it can accumulate in reservoirs thereby reducing the water supply or it may accumulate in estuaries and harbours and be a hazard to shipping.

Rain splash has now been recognised as an important erosive agent for at least two reasons. When rain drops hit the surface of the bare soil they cause material to be splashed upwards and outwards. This is of little effect on flat surfaces but on slopes some of the soil is splashed down the slope. A further effect of rain drop impact is that it puddles and seals the surface thus reducing infiltration which leads to increased run-off and potential erosion.

From the above it should be clear that soil erosion is bad because of the loss of valuable top soil and because of the problems it creates in reservoirs and the drainage system.

It should be mentioned also that soil is itself a reservoir for moisture. Some of the moisture from rainfall or melting snow is absorbed by the soil and percolates slowly through, eventually reaching streams and rivers, thus producing a fairly steady supply of moisture to the drainage. This is essential in order to maintain a constant domestic water supply, it also determines the nature of the fish-life in the river. If the moisture does not enter the soil but runs off over the surface after each shower of rain, then there will be wide fluctuations in the level of the rivers which could cause disastrous floods, leading to loss of life and property.

It is not possible to stop fluctuations in the level of most rivers, but sometimes they can be reduced considerably if run-off is controlled. In contrast, flooding has always been extremely beneficial in some areas. The annual flood of the Nile is virtually the only source of water for crop growth. Here the flood waters are carefully controlled to ensure that there is maximum infiltration and storage of moisture for the complete life-cycle of the crops.

Erosion control

The amount of run-off and soil erosion can be reduced by a number of methods. Probably the simplest procedure is to keep the surface constantly covered by vegetation. However, this is not possible when arable cultivation is conducted but it should always be practised on steep slopes, by having forests, orchard crops or permanent grass-land.

Wherever possible ploughing should be carried out along the contour so that there are no furrows down which water can run. A common form of erosion control is to construct terraces which restrict the speed of moisture movement as well as increasing greatly the infiltration on the flat terrace surface. The ancient terraces of many of the grape growing areas are world famous. This system not only reduces erosion but also stores moisture in the soil (Fig. 53). Another form of erosion control is strip cropping which consists of narrow strips of two or more crops grown alternately along the contour (Fig. 54). Usually each alternate strip is a biennial or perennial so that only alternate strips are bare at any one time, therefore any erosion that starts in the bare strip is checked by the crop on the next strip.

FIG. 53. The terrace cultivation of rice in Java

Fig. 54. Contour strip cropping in Wisconsin U.S.A.

Cultivation

With most systems of arable crop growth it is usual to cultivate the soil by the use of implements. The type and degree of cultivation will depend upon the type of crops that are grown as well as upon the economic status of the country. There are a number of reasons why the soil is cultivated. These include:

1. The production of an adequate depth of loose, porous top soil into which plants can freely extend their roots
2. The control of moisture, aeration and temperature
3. The destruction of pests and weeds
4. The mixing of plant remains and manures in the soil.

These ends can be achieved by a simple hand tool such as a hoe, by means of a primitive human or ox-drawn plough or by highly

Fig. 55. This Greek farmer rests his oxen

Fig. 56. A modern tractor/plough in action on a Scottish field

sophisticated tractor-drawn ploughs (Figs 55, 56). When growing grain crops such as wheat the soil is ploughed and then harrowed to produce an even surface and to break down the large clods. Seeds are then planted and at the same time fertilizer is added in such a manner that the concentration next to the seed is not too high, otherwise the germinating seedling may be killed or badly damaged.

If the crop is potatoes there may be further cultivation during the life-cycle of the plant. This will include weeding, ridging and a further application of fertilizer – a *top dressing*.

In addition to the factors discussed above there are a number of essential non-soil factors that plants require for adequate growth. These include solar radiation as a source of energy for photosynthesis, respiration and transpiration. Also required is optimum atmospheric temperature which varies from species to species.

Land use

Most of man's activities are concerned with the direct or indirect use of the land. As man developed and his technology improved he has made greater and greater demands upon the land so that more and more the potential of the soil is becoming a factor limiting his personal well-being, since the soil is required to produce more food and to carry more buildings. Thus it is steadily becoming the focal point in conservation and environmental studies.

Whereas early man cleared small areas of ground to create villages and to cultivate a few crops, modern man needs the ground for:

1. Growth of crops and animals for food on a permanent basis or as shifting cultivation
2. Building sites for factories and houses
3. Grazing the natural vegetation
4. Building roads and airfields. It is estimated that in Britain 20,000 hectares per annum are used for roads
5. Construction of playing fields for football, golf, horse-racing, etc.
6. Gravel pits and open-cast mining for coal, iron ore, etc.
7. Nature reserves and parks
8. Sewage disposal

9. Reservoirs, the water being used for power supply, domestic and industrial consumption and recreation.

Running in parallel with the growth in technology there has been an enormous growth in the world's population so that competition for the land has increased as a natural consequence. The nature of the competition is extremely complex, including rivalry between individuals for a given piece of land and the competition between man and the other creatures in nature. Thus we find around towns that there is competition for the use of land for factories, crops, housing or playing fields to mention only a few uses. Competition is not confined to the areas peripheral to towns, it occurs also in the heart of rural areas as in the uplands of Britain where there is strong competition between, forestry, sheep grazing and the shooting of deer and grouse. Throughout large parts of Australia one sees the very strong competition between sheep and the kangaroo for the relatively small amount of natural herbage. These various types of competition often lead to many types of misuse and damage to the land including the following:

1. Soil erosion due to deforestation, poor cultivation or death of vegetation caused by fumes from factories
2. Pest infestations such as potato eelworm due to continuous growth of potatoes
3. Pollution due to the addition of excess pesticides and herbicides
4. Pollution of the surface soil due to poor siting or over-loading of septic tanks. This can cause the spread of hepatitis
5. Salinisation due to poor cultivation and irrigation with saline water
6. Dumping of industrial waste creating slag heaps and the like.

Although the nature of the soil itself should be one of the major factors determining its use this may not be the case when economic considerations come first, often leading to misuse of the soil. Thus it may be possible to produce a land capability classification for a given area but implementing it might not be possible because of some specific local custom or the lack of capital. Probably taking the world as a whole it is ignorance or a low level of technology that is often responsible for the misuse or improper use of land. It is in countries with a high level of technology that one tends to find the best use of land and the highest level of production but there are notable exceptions to this in the form of peasant agriculture in many parts of Western Europe and catastrophic erosion in the U.S.A. and Australia. Thus at present we find similar soils being used in a variety of ways. In one area a given soil may have a well developed and balanced system of land use which has been built up following a soil survey. This would include agriculture, forestry and wild-life preservation, but in another area a similar soil may be supporting subsistence agriculture because of the lack of capital. In some countries efforts are being made to make good some of the damage that has been done, but in other cases it is too late, as in central Turkey where many hillsides have been eroded down to the bare rock. An excellent example of reclamation is in the Tennessee Valley in the U.S.A., where, by building dams and hydro-electric stations across the river, enough water and electric power are produced to change the standard of agriculture from subsistence to prosperity. This scheme plus others have shown that prosperity comes only when the utilisation of an area takes into consideration all the facets of the environment such as:

1. Soil erosion control
2. Maintenance of soil fertility by the addition of lime, fertilizers, drainage, etc.
3. Forest management and afforestation when needed
4. Wild life preservation
5. Elimination of pollution of the atmosphere and natural waters.

6

Soil maps and mapping

One of the major occupations of soil scientists is surveying and mapping soils, and the production of soil maps. Surveying soils is a most exacting scientific task because soils do not have sharp boundaries but gradually grade from one into another. Further, this gradation is seldom clearly expressed on the surface of the ground and has to be determined by making numerous examinations which take the form of small inspection pits or auger holes.

There are various methods for conducting soil surveys. Most are "free surveys" based on the experience of the soil surveyor who criss-crosses the area making inspections at points in the landscape where he considers it necessary to determine the nature of the soils and to establish boundaries. At each point of inspection the properties of the soil are recorded on a map and when necessary in a note-book. Alternatively, he carries out inspections at predetermined locations which are based on some type of statistical programme. This technique is more objective but is often time consuming.

The type of map on which the data are recorded can vary very much from place to place. In countries such as Britain, ordnance survey maps are used but for many places such maps have not been produced. Then it is customary to use aerial photographs and to record the data either directly on to the photograph or on to a transparent overlay. Aerial photographs have a number of advantages, particularly with regard to finding one's location but aerial photographs have a varying scale across the photograph therefore they cannot be used as complete substitutes for maps. The scale of the map or photograph varies depending upon the object of the investigation and time available. The scale seldom exceeds 10 cm to 1 km except for special purposes such as irrigation. From the data recorded on the map or photograph and in his note-book the surveyor draws lines on a map or photograph to enclose areas of relative uniformity and so to produce a soil map. The field map is then submitted to the drawing office for the preparation of the final coloured map and its accompanying legend.

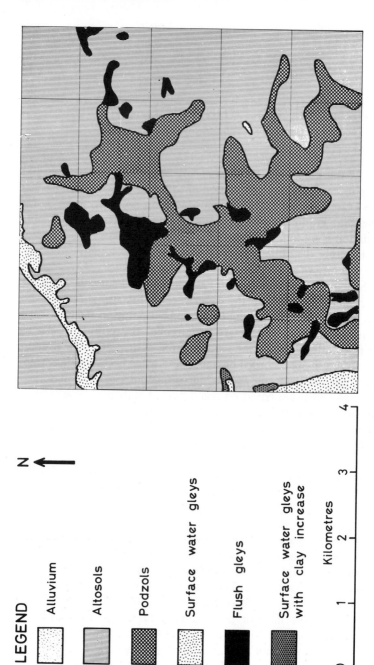

LEGEND

☐ Alluvium

▤ Altosols

▦ Podzols

░ Surface water gleys

■ Flush gleys

▨ Surface water gleys with clay increase

N

Kilometres
0   1   2   3   4

FIG. 57. Soil map of the area west of Church Stretton in Central England

FIG. 58.

Salt marsh and reclaimed land in Norfolk, England. Salt marshes are generally areas of silt accumulation which can be reclaimed by constructing a bank and a drain on the landward side of the bank, the material dug out of the drain being used to make the bank. The enclosed creek system is then rationalised, a few larger creeks are maintained as major drainage channels and the small ones filled in.

The distinctive pattern on the marsh is due to differences in the distribution of the natural vegetation, creek pattern and reclamation.

A. Dominated by the light coloured cordgrass with sea poa as the dark areas
B. Dominantly sea poa
C. Clearly defined well drained areas adjacent to the creeks and dominated by Sea-couch
D. The bank
E. The drain
F. The pattern of infilled creeks where cropgrowth is poor

When mapping on a large scale, it is possible to delimit the occurrence of each soil but as the scale gets smaller and smaller so do the areas on the map; then it becomes necessary to group soils that are in some way related. In some cases all the soils with the same type of parent material are grouped. In areas with a repeating topographic pattern the grouping is based on topography. This latter type of grouping which is sometimes referred to as a *land system* is very popular because a considerable amount of mapping can be based on aerial photography with a limited number of ground checks. With this system it is assumed that there is a correlation between topography, geology, vegetation and soils.

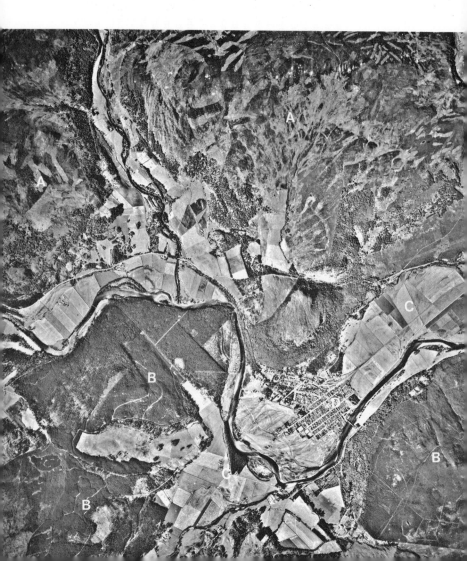

Accompanying the soil map is a report which gives further details about the soils particularly some of their chemical and physical characteristics. Fig. 57 is an extract from a much larger soil map. It shows the distribution of the soils and the spatial relationships between the various soils. The distribution pattern is determined largely by slope and elevation. The podzols occur >500m on a plateau surface. On the slopes leading down from the plateau there are altosols, while flush gleys occur mainly at the break of slope between the podzols and altosols. In the flat low-lying areas there are surface water gleys. In a number of reports there are diagrams which not only show the relationship between the soils but also relationships with slope and elevation. These are known as block diagrams such as shown in Figs. 4, 13, 20. The aerial photographs given in Figs. 58 and 59 are examples of areas where the different types of vegetation are strongly correlated with different soils. From the original soil map it is possible to derive a number of other maps and it is becoming customary to prepare a land capability map which is published together with the soil map and report. Although areas of soils can have a variety of uses, the land capability map is usually produced for agricultural purposes but they can be produced for special purposes such as housing or industrial development. Bibby & Mackney (1969) have produced a system of land use classification for the British Isles.

FIG. 59. (*facing page*)

Land use pattern around Ballater, Scotland. This is a good example of the wide range and strongly contrasting systems of landuse in Highland Britain
  A. The highest ground above 300m has a patchy pattern produced by burning the heather which regenerates to produce young shoots for grouse that are shot for pleasure
  B. At a lower level are forests dominated by Scots pine.
  C. The flat cultivated alluvial terraces form the lowest part of the landscape
  D. The town of Ballater is situated on alluvium and partly surrounded by a loop in the river Dee

Soils are the world's major natural resource and a soil map is the spatial representation of these resources. Therefore soil maps are fundamental and should be the starting point when planning every type of land use whether it be for agriculture, forestry, road building or house construction. It is essential that the land surface should be utilised to its optimum, for example, good agricultural land should be kept for agriculture and not used for house building. A particularly striking example of the misuse of land occurs near London where Heathrow Airport was constructed on some of the best soils in Britain.

7

Geography of world soils

An account of the major soils of the world should be preceded by a discussion about soil classification but this has been purposely omitted because of the lack of agreement among soil scientists about a single method of classifying soils. Thus the soils of the world are presented in a general way using as far as possible the classical pedological terminology but in a few cases modern terms are used in order to avoid the perpetuation of some ambiguities (FitzPatrick, 1971).

The soils are presented mainly in the form of a transect; the first part is down through the U.S.S.R. west of the Ural mountains and the second from the Sahara desert to Zaire. This treatment illustrates the broad geographical relationships between soils in two of the main continental areas of the world (Fig. 60). In addition a few important soils are mentioned that form as a result of maritime conditions or resulting from the local dominance of a particular soil forming factor.

In the north of the U.S.S.R. where conditions are very cold one finds **tundra soils** (Colour Plate IA and Fig. 68) which are characterised by a thin accumulation of organic matter at their surface followed by a dark greyish brown mixture of organic and mineral material below which is a wet mottled horizon about 50 cm thick. Then there is a sharp change into the permanently frozen subsoil or permafrost composed of alternating lenses of pure clear ice and frozen soil. The upper horizons freeze every winter and thaw every summer, a cyclic process which causes expansion and contraction and the formation of surface patterns such as mud polygons, stone polygons (Figs. 25, 26) and tundra polygons, which are often outlined by small depressions beneath which are vertical wedges of ice extending to a depth of 3 m or over (Fig. 61). The range of plant species in the tundra is restricted to a few mosses, liverworts, lichens, grasses and sedges but to the south with higher temperatures, there are shrubs such as the arctic willow and finally spruce and larch forests.

Tundra soils are useless for agriculture because of the low temperatures, but they support large herds of reindeer and caribou which either roam freely or are crudely managed. At present these herds face two major hazards: they are often indiscriminately hunted and the lichens that they eat accumulate strontium from radio active fallout causing a decline in the fertility of the herds.

South of the tundra there are extensive areas of **peat** (Figs. 60 and 69) with its marshy vegetation dominated by *Juncus* spp., *Carex* spp.

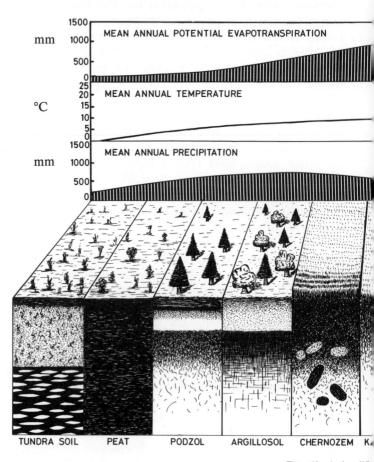

Fig. 60. A simplifie

and mosses, particularly *Sphagnum* spp. Peat is the accumulation of organic matter under wet anaerobic conditions and is divided into basin peat and climatic peat. Basin peat forms in wet depressions and flat sites where the plant litter will not decompose rapidly but accumulates. Although the rate of decomposition of peat is slow, some changes do take place so that it ranges from very fibrous and woody to amorphous and plastic. Basin peat occurs elsewhere as in the Everglades of Florida and on the north coast of Borneo (Figs. 63 and 64).

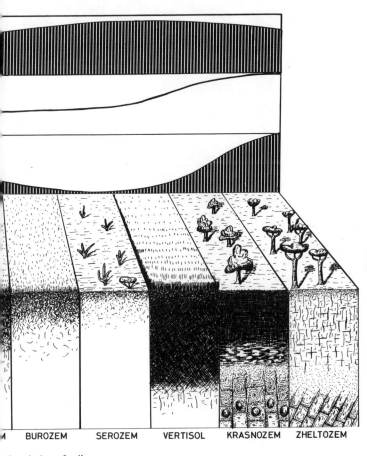

M   BUROZEM   SEROZEM   VERTISOL   KRASNOZEM   ZHELTOZEM

al variation of soils

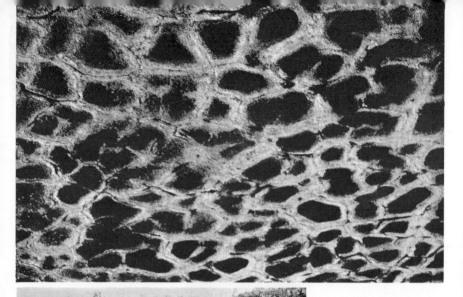

Oblique aerial view of
tundra polygons in
northern Alaska. Each
polygon is 15–20m in
diameter

FIG. 61B.

An ice wedge 4m deep on
Gary Island, northern
Canada

Climatic peat forms in areas of high atmospheric humidity, low evapotranspiration and consequently wet soils. These conditions are found on the humid west coasts of Canada, Alaska, Britain and Scandinavia.

Basin peat receives much of its moisture by run-off which varies in composition and acidity. The fens of England are a good example of

FIG. 62. Fossil ice wedge in Scotland. The vertical wedge which is 3 metres deep is composed of sand and gravel but was solid ice when there was a tundra climate

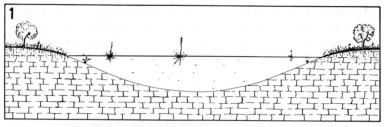

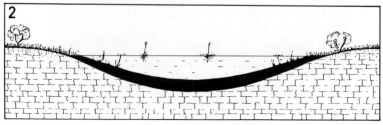

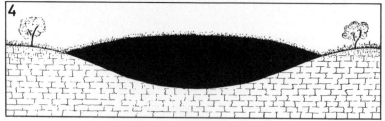

Fig. 63. Peat formation

1. A small pond or lake which could be situated between two moraines.
2. At the bottom of the pond there is a thin accumulation of organic matter from the plants growing in the pond and in the surrounding soil.
3. Considerable increase in the thickness of the organic matter and the spread of vegetation on to its surface.
4. Continued thickening of the peat to develop the characteristic domed form of the final stage.

peat with high pH values due to the high content of calcium carbonate in the run-off water derived from the neighbouring chalk.
On the other hand climatic peat is always acid because it derives its
moisture from the atmosphere which contains very little basic
cations.

Peat is cultivated extensively, but often needs to be drained and
must be limed to raise its pH (except for fen peats). In addition it
often suffers from copper and zinc deficiency. The precise type of
crop varies from place to place and include forests, permanent grassland and crop plants such as onions.

Peat is the traditional fuel in many parts of the world but it is
more valuable for horticulture.

FIG. 64. A peat deposit containing a layer of tree stumps that
indicates a dry phase and forest cover during its accumulation

Great areas of coniferous forests follow the peat and here the soils are predominantly **podzols** (Colour Plate IB and Fig. 70) on the freely draining sites with the morphology and processes mentioned in Chapter 1. There is a litter layer followed by progressively more decomposed organic matter then an organic mineral mixture, a bleached sandy horizon, a horizon of sesquioxide accumulation and finally the parent material. The water percolating through the decomposing organic matter dissolves acids then it enters the mineral soil causing decomposition of the minerals and the release of ions such as calcium, potassium and iron. These are translocated downwards leaving behind the light coloured upper horizon and the brown coloured middle horizon of sesquioxide and humic desposition. These soils are usually very acid, with pH values of < 4·5 in the surface increasing to about 5·5 in the lowest horizons. Also they are very deficient in plant nutrient ions except in the surface organic matter which releases them upon decomposition. These ions are quickly taken up again by the plants thus forming a cyclic process.

Podzols are inherently infertile for arable crops. Thus in order to cultivate them, tree stumps must be removed followed by liming and ploughing. This raises the pH and mixes the soil to encourage the growth of bacteria which break down the organic matter. Surprisingly, they are now amongst the world's most productive soils and support a wide range of crops and various types of animal husbandry. This is because they occur mainly in countries that have a very advanced technology. Because podzols are very sandy soils, crop growth on them is similar to sand culture since a constant addition of lime and fertilizers is necessary.

Associated with podzols in undulating topography are gley soils and peat which usually occur in the depressions (Fig. 65).

Gley soils are divided into two groups, the **subgleysols** (Plate IIA) (ground water gleys) and the **supragleysols** (the surface water gleys). The subgleysols form when the water-table is near to the surface. They have organic matter at the surface followed by a dark organic-mineral mixture then a grey or olive coloured horizon with marked yellow and brown mottling. This grades with depth into a completely grey, olive or blue horizon. The mottled horizon forms due to alternating wet and dry periods. When the soil is wet and anaerobic the iron is reduced to the ferrous state but during the dry period air is drawn into the soil and there is local oxidation along cracks and old root runs to give yellow and brown colours. The lowest horizon is

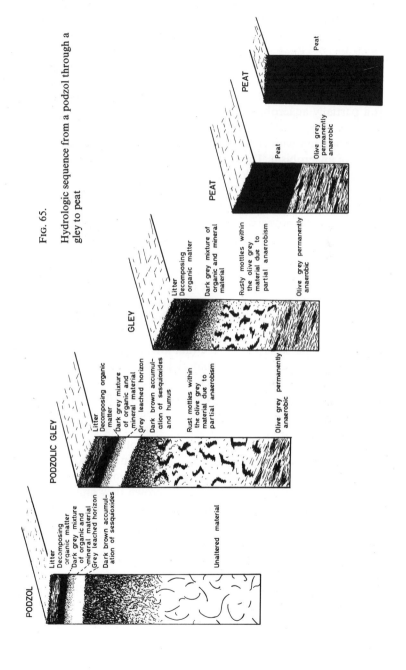

Fig. 65.

Hydrologic sequence from a podzol through a gley to peat

constantly anaerobic hence the grey-blue or olive colours. Subgleysols usually have a plant community with wet habitat species which.may include alder and sphagnum. Supragleysols form because the soil has a slow rate of percolation so that water tends to accumulate within the middle part of the soil. They also have a wet habitat vegetation. They have litter and partially decomposed organic matter followed by a dark organic mineral mixture. Below is a partially anaerobic, grey, olive or pale brown, faintly mottled horizon. This usually changes sharply into a brown or reddish brown horizon with some mottling, but generally there is less evidence of anaerobism. These soils have a fine texture or there may be an increase of clay with depth causing the reduced permeability and the water to accumulate. Thus supragleysols have their horizon of maximum reduction sandwiched between two aerobic horizons.

The principal property limiting the utilisation of gley soils is their wetness, therefore they have to be drained, then they can be used for producing a wide range of crops. As stated above (Fig. 65) there is a topographic relationship between podzols, subgleysols and peat.

Thin iron pan soils (Colour Plate IIB), distinguished by the presence of a thin continuous iron pan, are very common in Britain and Scandinavia. They usually have a heath or moorland community dominated by various *Ericaceous* species, *Carices*, lichens and mosses. At their surface there is a thin litter followed by up to more than 20 cm of well humified material. Then there is an organic mineral mixture followed by an olive grey faintly mottled horizon, underlain by the thin hard continuous iron pan, which may rest on a variety of horizons but they are often strongly compacted.

The iron pan is impervious, so that water is held up in the upper part of the soil, causing the formation of the olive grey horizon and the accumulation of organic matter. In addition these soils are very acid and nutrient deficient which render them almost useless for any type of utilisation. Before crops can be grown they must be limed, fertilized and cultivated, but above all the iron pan must be broken to improve drainage. This last operation requires expensive equipment which is not often available. At present considerable efforts are being made to establish coniferous forest on these soils, particularly in the British Isles.

The very fertile **altosols** (Colour Plate IIIA and Fig. 3) occur beneath some deciduous woodland of the cool temperate areas. At the surface there is a loose leafy litter resting on a brown granular horizon containing numerous earthworms. Below is a brown, friable, middle horizon which grades into the parent material that is often basic or calcareous, imparting a high base saturation and slightly acid to mildly alkaline pH values. These properties encourage earthworm and bacterial growth thus causing a rapid breakdown and incorporation of organic matter into the soil. Thus these soils have a high natural fertility and the progress of early man through many parts of central Europe was such that he first cleared the forests from altosols and cultivated them and only when the pressure on the land became high did he turn his attention to the less fertile soils such as podzols. The forests were oak and beech which were very valuable as building materials so that altosols provided the timber for his home and a suitable seed bed for his crops. However, altosols need lime and fertilizers to support continuous cultivation.

There is a continuous graduational sequence from altosols to podzols in which the soils become progressively more acid and leached. In some cases altosols can be changed into podzols by planting conifers on them.

In many cool temperate areas of the world with deciduous forests or mixed deciduous and coniferous forests there are **argillosols** (Colour Plate IIIb). These soils have a middle horizon containing more clay than the horizons above or below, as a result of translocation of clay from the upper horizons. Thus the profile of argillosols has a thin loose leafy litter resting on a greyish brown mixture of organic and mineral material. This changes into a grey sandy horizon followed by a sharp change into the brown blocky or prismatic horizon which has the high content of clay, then there is a gradation to the parent material.

Argillosols are moderately acid at the surface with low base saturation which imparts fairly low natural fertility thus they need lime and fertilizers as a normal part of arable cultivation. At present they are very productive, supporting a wide range of crops and various types of animal husbandry, particularly dairying, because they occur in areas with a high level of technology. It appears that soil with the maximum clay in the middle horizon can form by a variety of other processes. Two other processes are the destruction and removal of clay from the surface and maximum weathering in the middle horizon, forming **luvosols** and **clamosols** respectively.

South of the deciduous forests precipitation falls to less than 400 mm and there is a zone of tall grass growing on **chernozems**

(Colour Plate IVA and Fig. 71) which is the Russian for black soils. These soils have a root mat at the surface resting on a thick black horizon up to 2m in thickness. Within the lower part of this horizon and below there are thin thread-like deposits (pseudomycelium) of calcium carbonate and also there may be concretions of calcium carbonate. The black horizon grades into the yellowish brown parent material which is usually loess.

A conspicuous feature of European chernozems is the presence of burrows (crotovinas) caused principally by the blind mole rat. These crotovinas extend into the parent material and are seen as dark infillings in the lighter coloured parent material or as light areas in the dark top soil. In addition these soils have a vigorous earthworm population. Because precipitation is <400mm, only enough water passes through the soil to remove the most soluble salts and to translocate calcium carbonate, hence the pseudo-mycelium and concretions. These soils have pH values about neutrality, very high base saturation, and a very high inherent fertility. They form the main areas of grain production, particularly wheat and more recently maize. Initially it was thought that they had an inexhaustible fertility but after decades of cultivation, production has gradually declined so that it is now customary to add fertilizers.

In the U.S.A. and elsewhere there are **prairie soils** or **brunizems** which are similar to chernozems but they have a middle horizon with a clay maximum and are slightly less fertile.

South of the chernozems the precipitation gradually decreases from 400mm to 100mm and the vegetation changes from tall grass, to short grass, to bunchy grass and then into species that will with-stand long dry periods. There are parallel changes in the soils and gradually the thick black horizon becomes lighter in colour and shallower and the horizon of calcium carbonate comes nearer to the surface but generally three distinct soil types can be recognised, viz., kastanozems (chestnut soils), burozems (brown soils) and serozems (grey soils). **Kastanozems** have a short grass vegetation and dark brown horizon up to about 30cm thick, below which is the horizon of calcium carbonate accumulation followed by the unaltered material. **Burozems** have a bunchy grass vegetation and a thinner brown upper horizon then the lower calcium carbonate horizon. **Serozems** have typical desert type vegetation and a thin brownish grey upper horizon followed by that of carbonate accumulation.

These three soils have a very high inherent fertility but productivity is usually restricted by the lack of moisture. This can be overcome by irrigation or moisture conservation but often in these areas there are only a few rivers or other sources of irrigation water. An exception is the Colorado river which runs through the desert in the south western part of the U.S.A.

When irrigation is impossible dry-farming is carried out particularly in areas of burozems. In areas of serozems water conservation by dry-farming is inadequate to support crop growth, irrigation is essential.

Within the semi-arid areas where evapotranspiration greatly exceeds precipitation, certain cations and anions accumulate in the soils to cause high salinity or alkalinity. The ions include sodium, potassium, magnesium, calcium, chloride, sulphate, carbonate and bicarbonate. Since the upper limit for salt tolerance by most plants is about 0·5 per cent, soils containing higher amounts are regarded as being saline and generally referred to as **solonchaks** and are usually easily recognised by salt efflorescences on their surfaces (Fig. 44). The soil profile often resembles a subgleysol soil by having an upper grey organic mineral mixture which rests on a mottled horizon followed by a grey or olive completely reduced horizon.

Many solonchaks are potentially useful for agriculture if the excess ions can be dissolved and removed. This requires large volumes of water but sometimes a suitable supply is not available because of the general lack of adequate amounts of salt-free water in these dry areas.

In some soils of dry areas a water-table with a large amount of dissolved salts may be present at depth. If such soils are over irrigated, as is often the case, the irrigation water reaches down to the water table and may cause it to rise to the surface by capillarity, thereby inducing salinity and severely reducing crop production. This has happened in many areas, a good example being the Sind valley in Pakistan where thousands of hectares are lost to agriculture every year (Fig. 66).

Solonetzes also occur in semi-arid areas and have at their surface a thin litter followed by a thin very dark mixture of organic and mineral material and then a dark grey somewhat sandy horizon. Below is the very distinctive middle horizon with its marked clay increase, characteristic prismatic or columnar structure and a pH value that is often over 8·5 due to high exchangeable sodium and magnesium but generally a low content of salt (Fig. 40 and 41). The high pH is extremely harmful to plant growth and must be lowered

FIG. 66. Soil spreading in farmland in Pakistan. The inset shows the salt crust that forms part of

before cultivation can be carried out. This is usually achieved by adding calcium sulphate, the calcium enters the exchange complex, replacing sodium which is removed by leaching following natural rainfall or irrigation. The calcium has other beneficial effects for it improves the structure and is an essential element for plant growth and microbiological activity.

Solods can be regarded as leached solonetzes in which the upper horizons are strongly bleached becoming pale grey or white. The middle horizon has a clay maximum, high exchangeable sodium and/or magnesium, but it is acid. When adequate amelioration has been carried out these soils can be used for growing a wide variety of crops.

Desert soils occupy about one fifth of the earth's surface and in the Sahara they are rocky, stony or in the form of sand dunes. Also there are areas of loose material which is not really soil but they have a high potential for crop growth if irrigated. However, the intense solar radiation causes plants grown under these conditions to transpire very rapidly during the day, leading to physiological drought and often permanent damage. Therefore, desert areas in many cases offer a low potential for crop production.

Within semi-arid regions of the tropics and sub-tropics there are extensive flat areas with deep dark coloured clay soils generally known as **vertisols** (Fig. 72). The surface horizon is usually granular but can be massive and is followed by a dense horizon with prismatic or angular blocky structure that grades into similar material with a marked wedge structure. This is caused by pressures developed as a result of expansion and contraction of the montmorillonite clay in response to wetting and drying. Many of the peds have slickensides i.e. shiny surfaces which form as one ped slips over the other during expansion (Fig. 27).

Vertisols usually have pH values about neutrality, a high base saturation and high C.E.C., therefore they have a high potential for agriculture but often suffer from drought; but under irrigation their productivity is very high. Perhaps one of the most famous areas is the Gezira of the Sudan where there is a high output of excellent quality cotton.

Within the humid tropics occur bright red soils known as **krasno-zems** (Colour Plate IVB and Fig. 73). These soils have a thin litter at the surface followed by a greyish red mixture of organic and mineral material which is not more than 5–10 cm thick. This grades quickly into a bright red horizon which may be several metres thick followed by a change into the underlying rock. This red horizon is usually composed largely of kaolinite with some iron oxides but there may be a very high content of gibbsite. It is then regarded commercially as bauxite and mined for the production of aluminium. The change to the underlying rock can be gradual through progressively less weathered rocks or it may be sharp. Often there is a characteristic red and cream mottled horizon beneath the red top soil or there may be a thick white horizon known as the pallid zone. In many of these soils there is a vesicular slag-like horizon generally known as laterite which may be soft within the soil but hardens upon exposure. In West Africa it is very common for a horizon of gravel accumulation and a stone line to occur just beneath the surface. A third variation is the occurrence of a horizon composed largely of concretions which may vary in size from 1 to more than 50 mm in diameter.

Krasnozems are formed by the progressive weathering of the rock until all of the primary minerals such as feldspars, amphiboles and pyroxenes have been completely decomposed and transformed into secondary substances or lost from the system, hence the fact that the middle horizon has a high content of kaolinite or oxides (Fig. 34). Because weathering has been so complete these soils are very deficient in essential plant elements but they often carry very luxuriant high forest because the elements are constantly being recycled and there is a small reservoir in the top few centimetres due to their release by the decomposing litter. This caused many earlier workers in the tropics to be misled for they assumed that the high forest was indicative of a very rich soil. In the majority of cases when the high forest is cut and agriculture attempted there is failure within two or three years, during which the essential plant elements in the surface are exhausted. This was known to the native people who practised a system of shifting cultivation. They would clear an area and cultivate it for two or three years and then move on to a fresh site allowing the depleted site to develop a secondary forest and for the reservoir of fertility to be rebuilt. When fertilizers are available these soils respond well to cultivation and grow a variety of crops such as cocoa, coffee, sugar cane and oil palm.

Because of the wide climatic fluctuations during the Pleistocene period many krazsnozems that developed under humid tropical conditions are now in another environment. This is particularly marked in Australia where red soils and laterite occur commonly in the desert and elsewhere.

Although krasnozems are common soils within the humid tropics there are other soils that are shallow and not as strongly weathered and contain primary minerals (**rufosols**). Also, some of the soils that occur under moister conditions are yellow and brown and are known as zheltozems.

Some of the most dramatic changes brought about by man are seen in some tropical and subtropical countries as a result of growing rice. Many brightly coloured soils now have drab and mottled colours due to the long periods of waterlogging required by the rice plants.

In a number of tropical and subtropical areas with a marked dry season such as Western Australia, there are distinctive yellow and yellowish red soils containing an abundance of subspherical concretions about 1 cm in diameter. These soils are about 3 m thick and are usually loose at the surface but may be cemented at a depth of 1–2 m. Below is the zone in which the concretions are forming and overlying the weathered rock (pallid zone). Although these soils are very common in some places they have not been named, nor have the processes of formation been studied. Because the dominant property seems to be the presence of subspherical concretions they might be called **spherosols** (Fig. 67).

Arenosols are composed predominantly of coarse sand dominated by quartz and occur in many tropical and subtropical areas with a marked dry season. They may be several metres thick with uniform colour and range from bright red to yellow and even pale brown or grey depending upon climatic conditions. They can develop from sandy parent material or they may form by the differential removal of fine material leaving a concentration of sand behind. These soils are inherently infertile. However, they can produce a number of crops if fertilizers are applied but moisture is often the principal limiting factor.

Andosols are formed mainly on volcanic ash. At the surface is a loose litter followed by a dark coloured mixture of organic and mineral material which changes fairly sharply into a brown middle horizon that grades into the parent material. The unique property of these soils is their high content of allophane which gives them a very low bulk density and fluffiness particularly in the middle horizon. Probably the greatest extent of these soils is in Japan and New Zealand.

FIG. 67. Concretions in a spherosol

Two distinctive soils that develop from limestone are **rendzinas** and **terra rossas**. Limestone is an exceptional rock, being composed largely of soluble calcium carbonate which is dissolved during weathering so that only the very small amount of impurities in the rock provide the residue to form soils, which as a consequence are never very deep.

Rendzinas have an upper black or very dark brown granular horizon speckled with white fragments of limestone. This usually does not exceed 50 cm and may change abruptly into the white limestone or there may be a narrow transition composed of the

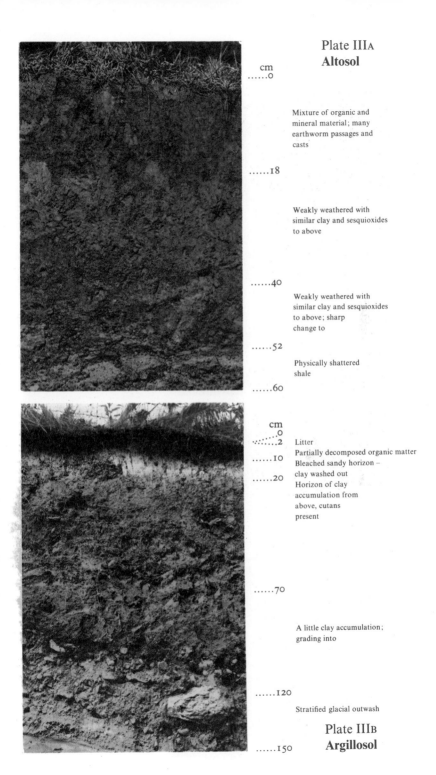

Plate IIIA
Altosol

cm
......0

Mixture of organic and
mineral material; many
earthworm passages and
casts

......18

Weakly weathered with
similar clay and sesquioxides
to above

......40

Weakly weathered with
similar clay and sesquioxides
to above; sharp
change to

......52

Physically shattered
shale

......60

cm
......0
......2   Litter
         Partially decomposed organic matter
......10 Bleached sandy horizon –
         clay washed out
......20 Horizon of clay
         accumulation from
         above, cutans
         present

......70

A little clay accumulation;
grading into

......120

Stratified glacial outwash

Plate IIIB
Argillosol

......150

Plate IVA
Chernozem

cm
......0

Mixture of organic and mineral material, abundant earthworm passages and casts, numerous passages caused by the blind mole rat, pseudomycelium of calcium carbonate near the base

......85

Abundant, recent and old crotovinas, pseudomycelium and concretions of calcium carbonate

......150

Loess parent material with frequent vertical earthworm passages

......270

Rain forest

Cultivated grassland

cm
......0

Clay with well developed angular blocky structure, may remain uniform for several metres then grades gradually into weathered rock underlain by the solid rock

Plate IVB
Krasnozem

......150

black upper horizon and larger fragments of limestone. These soils are usually extremely fertile with pH values about neutrality, but may suffer from drought during dry years because of their small volume for water storage (see Fig. 45). For this reason many rendzinas are not cultivated but have a natural or semi-natural vegetation such as the beechwoods of the Chiltern Hills in Southern England and numerous parts of the Mediterranean.

Terra rossas are characteristically brilliant red in colour and commonly are more than one metre in thickness. Terra rossas are very old soils and represent the end stage of soil formation on limestone during which all of the carbonate has been removed, just leaving behind the accumulation of impurities and can be regarded as a type of krasnozem. They occur in many humid parts of the world but they are principally associated with the Mediterranean where they are used extensively for growing grapes. This area, having been cultivated since prehistoric times, means that terra rossas are probably among the world's most eroded soils – a feature that was noted by Plato.

Alluvial soils. Bordering most of the world's major rivers are soils developed in deposits of alluvium. In a number of cases these soils are extremely fertile and sustain a very high level of agriculture. It was upon the alluvial soils of the Nile, Tigris and Euphrates rivers that the first major western civilisations developed. This is not a mere coincidence for without fertile soils and an abundant supply of food, healthy and thriving communities cannot develop.

Most soils developed in recent alluvium are either primosols or rankers (see pages 49 and 52). Therefore they represent the initial stages in the formation of a soil. For this reason many workers consider the term "alluvial soil" as inappropriate.

Although the transects through other continental land masses such as North America sometimes give similar sequences there are often major differences. For example in Australia deeply weathered rocks can be found throughout most of the country and within the most arid parts of the interior. This weathering probably took place during a much wetter climatic phase and with a change to aridity the soils have many characteristics of those found in a more humid environment. Thus they display a unique polygenesis.

GEOGRAPHY OF WORLD SOILS

The series of distribution maps on pp 125–130
contains the following:

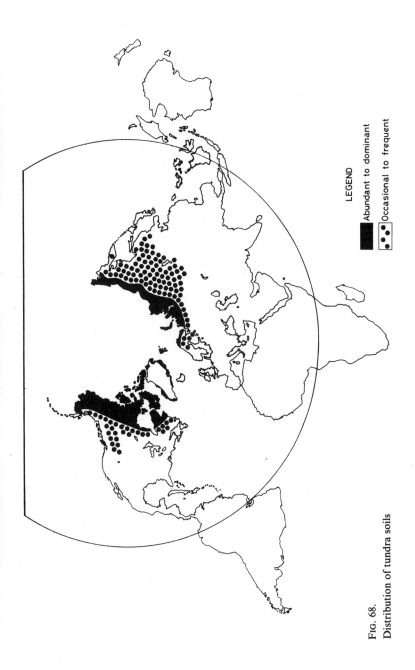

LEGEND

■ Abundant to dominant

▨ Occasional to frequent

Fig. 68.
Distribution of tundra soils

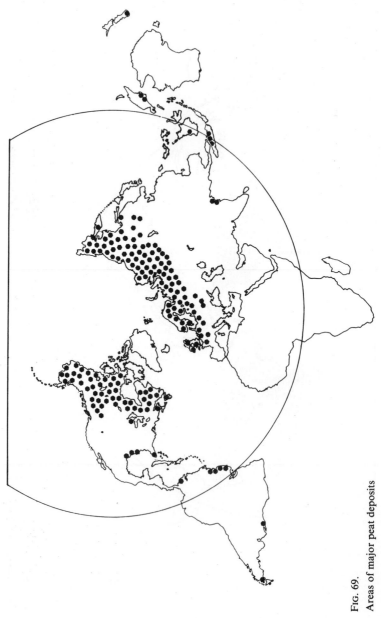

Fig. 69.
Areas of major peat deposits

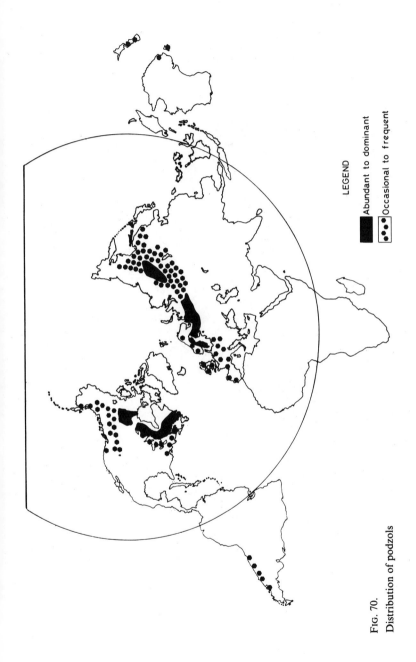

Fig. 70.

Distribution of podzols

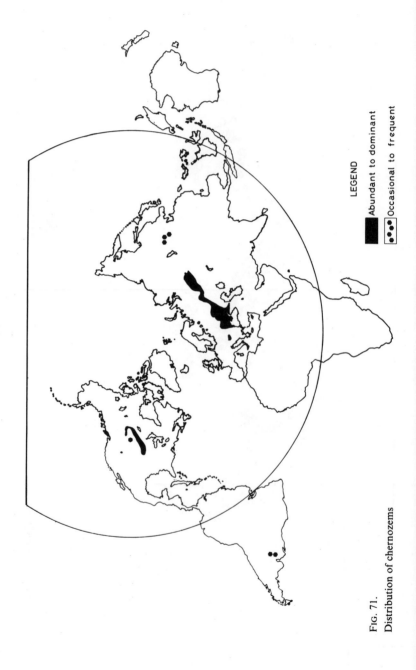

LEGEND

Abundant to dominant

•••• Occasional to frequent

Fig. 71.
Distribution of chernozems

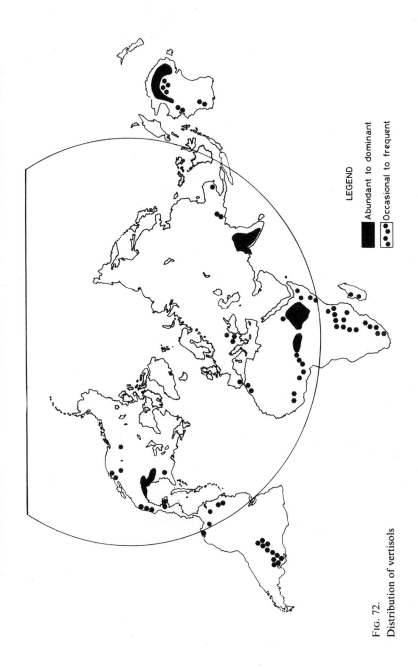

LEGEND

Abundant to dominant

Occasional to frequent

Fig. 72.
Distribution of vertisols

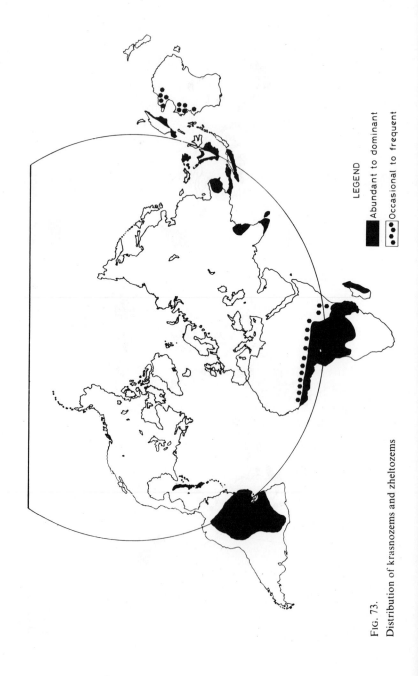

Fig. 73.

Distribution of krasnozems and zheltozems

Appendix I. Additional teaching material

Films

The making of soil.

Central Office of Information,
Herculese Road,
Westminster Bridge Road,
LONDON S.E.1

Getting down to drainage.

Central Film Library,
Government Building,
Bromyard Avenue,
Acton,
LONDON, W37 JB

The precious soil.

Produced by ICI Agricultural Division Film Unit,
Billingham, Tees-side.
Directed by Gerald Fisher.
Photography: Roger Thomas.
Scientific Adviser: A. J. Low.

Available for purchase or free loan from:
ICI Ltd.,
Central Film Library Thames House,
North Millbank,
LONDON, SW1P 4QG

Transparencies

The following transparencies can be obtained directly
from the author:

> Dr. E. A. FitzPatrick,
> Soil Science Department,
> University of Aberdeen,
> Meston Walk,
> ABERDEEN, AB9 2UE

in colour:

Tundra Soil	Solod
Podzol	Vertisol
Subgleysol	Cutans
Thin iron pan soil	Supragleysol
Altosol	Zheltozem
Argillosol	Spherosol
Chernozem	Gilgai
Krasnozem	

line diagrams:

A podzol profile
An altosol profile
Utilization of solar radiation
The moisture cycle
The utilization of precipitation
Life in the soil
The carbon cycle
The nitrogen cycle
The soil plant system
Latitudinal sequence of soils

Appendix II. Practical exercises

The only way to gain a thorough understanding of soils is to carry out investigations both in the field and in the laboratory, particularly the former; therefore frequent visits should be made to the field to see various types of soils, their factors of formation and their utilisation. The practical investigations of soils should start by examining a profile in the field and collecting samples. If possible, choose an area at which Exercise 1 and a number of the projects might be conducted. If the soil studies are paralleled by studies in ecology, botany or zoology then a site should be chosen so that the field work in all the classes can be conducted at the same place. A low hill with woodland adjacent to agricultural fields might be very satisfactory. This will allow samples to be collected from the crest to the bottom of the hill and for a contrast to be made between a woodland and a cultivated soil. For dwellers in large towns it may not be possible to get a rural site, but it should be possible to get a site within a park or botanical gardens.

It is assumed that these exercises will in most cases be carried out under the supervision of an experienced person who will advise about where to collect samples and the organisation of the work. The exercises are:

1. Examination of a soil profile
2. Collecting a range of soils
3. Soil sampling in towns
4. Preparing the samples for analysis
5. Examination of soil with the binocular microscope
6. Chemical composition of soils
7. Chemical composition of plants – dried grass
8. Release of ions by weathering
9. Examination of soil minerals
10. Heat movement in soils
11. Estimation of organic matter in soils

12. Demonstration of the presence of living organisms in soils
13. Demonstration of fungi in soils
14. Extraction of some fauna from fresh soil with the Tullgren funnel
15. Estimation of earthworm population
16. Estimation of texture
17. Investigation of particle size distribution
18. Water retaining capacity of soils
19. Determination of pH
20. Determination of the lime requirement
21. Demonstration of exchange properties

Exercise 1 — Examination of a soil profile

Equipment

Spade	Tape measure
Pick	Note-book or description forms
Trowel	Polythene or paper bags
Pencil	Trays or polythene sheets

Procedure

Dig a soil pit at a site chosen by oneself or by the supervisor but in either case permission must be obtained from the landowner. Generally the pit is about $2 \times 1\,m$ at the top and extends down to unaltered underlying material which usually occurs at over $1\,m$. Usually the pit has to be filled after the exercise is complete, therefore make two neat piles of the material from the pit. One pile contains the turfs and top soil and the other contains the subsoil, if possible place the soil on polythene or canvas sheets. Orientate the face to be examined so that it has maximum illumination, taking great care to ensure that the surface above the face to be examined is not tramped upon.

When the pit has been dug, carefully clean the face to be examined with a trowel. Now delimit the horizons and remove lumps of soil from each horizon and describe the properties given below for each horizon.

Depth in centimetres

Colour in general terms; professional soil scientists have special colour comparison charts

Texture (see Exercise 16)

Stones – size, shape and frequency

Structure (see page 61)

Handling properties – place a lump in the palm of the hand and gradually, close the hand observing whether the material is loose, friable, firm, hard or plastic. Sands are loose, loams are friable, clay loams are firm and clays are hard or plastic

Organic matter – amount and distribution

Roots – size and frequency

Water conditions – moist, damp, wet, waterlogged.

A form similar to that given on page 136 can be used for recording the data. In addition to the soil properties many of the environmental factors have also to be recorded.

After the description has been completed collect samples in polythene bags for laboratory examinations. The amount and type will depend upon the nature of the course. When conducting a full and detailed course, collect samples of about 2 kg each from each horizon. If the class is more general, it may be necessary only to collect a sample from a single horizon which would be the topmost mineral layer. Now refill the pit by putting the subsoil in first then followed by the topsoils and turfs. Take the samples to the laboratory and place them into trays or on to sheets of polythene, allow to dry and prepare them as described in Exercise 4.

Exercise 2 – Collecting a range of soils

Soils vary considerably from place to place so that in order to appreciate this variability many of the following exercises should be conducted on a range of soils. These may form part of a teaching collection and be available in the laboratory but it is more interesting to collect the samples oneself.

Equipment

Spade

Small polythene bags

Large bags – empty fertilizer bags that have been thoroughly cleaned are very suitable

Procedure

Collect from 4–6 contrasting sites enough soil from the top 15 cm in the large bags to give each participant 2 kg from each site. If

TABLE II: *Soil Description*

DATE:	PROFILE No:	LOCALITY: Map reference:	CLASSIFICATION:
PARENT MATERIAL AND AGE OF SITE	CLIMATE Precipitation: Temperature:	TOPOGRAPHY Elevation: Slope: Aspect: Exposure:	VEGETATION: LAND USE:
DRAINAGE Site: Soil:	EROSION	ROCK OUTCROPS	

Drawing of Profile	Depth (cm)	Horizon No. or Name	Colour	Texture	Stones	Structure and Porosity	Handling Properties	Organic Matter	Roots	Water Conditions

possible the range should include a sand, a loam, a clay loam, a clay and a calcareous soil. In addition, try to get samples of ground rock such as granite, basalt, slate and limestone. Take the samples to the laboratory and place them into trays or on to sheets of polythene, allow to dry and prepare them as described in Exercise 4.

Also collect 4–6 10cm cubes of soil from each site and place each in a labelled polythene bag for Exercise 5. Try to keep these cubes undisturbed.

Exercise 3 – Soil sampling in towns

Equipment
Trowel
Polythene or paper bags

Procedure

Although it may not be possible to examine soil profiles in towns, it is usually possible to get a satisfactory sample on which to perform laboratory examinations. The sample may be obtained from one's garden or from a nearby park, after getting the keeper's permission. In either case, collect about a 2kg sample of soil from the top 15cm, place it into a labelled polythene bag, take it to the laboratory, place it in a tray or spread it on to a sheet of polythene, allow to dry and prepare as described in Exercise 4. Also collect a 10cm cube of soil and place it in a labelled polythene bag and try to keep it undisturbed for Exercise 5.

Exercise 4 – Preparing the samples for analysis

Equipment
Wooden mortar and pestle
2mm sieve 10–20cm diameter
2kg jars or polythene bags

Procedure

When the soil samples are dry, carefully crush each with wooden mortar and pestle and sieve through a 2mm sieve. Weigh both the material on the sieve and that passing through. Discard the material >2mm (or it may be kept for Project 4) but store the material <2mm in a dry well stoppered jar or polythene bag.

Calculate the amount of each fraction as a percentage of the total.

Exercise 5 – Examination of soil with the binocular microscope

Equipment
Binocular microscope with incident light
Dissecting needle.

Procedure

Take one of the cubes from Exercise 2 or 3 and very gently break it open to expose the undisturbed inner part. Place one portion on to the stage of the microscope and with the aid of the dissecting needle carefully examine the freshly exposed surface. Note the extent to which aggregates have formed and the way in which the roots grow through or around the aggregates. In some soils there will be fungal mycelium or small arthropods. Place a drop of water on to the soil surface and carefully observe what happens. In most cases it will disappear into the soil showing that it is porous. Make a full description of your observations, illustrated where possible with drawings. Repeat the observations on the cubes from the other sites.

Exercise 6 – Chemical composition of soils

This exercise together with Exercise 7 show that soils and plants contain a number of different elements but the proportions in each are very different.

Equipment

50 ml measuring cylinder
400 ml beaker
Filter funnel
Gauze
Tripod

Bunsen burner
Filter paper
Glass rod
Test tubes in a rack

Reagents

2M hydrochloric acid
Usual test reagents for iron, aluminium, calcium, magnesium, potassium, sodium, phosphorus

Procedure

Place 50 ml of a non-calcareous soil into a 400 ml beaker and add

100 ml of 2M hydrochloric acid. Place the beaker on a gauze resting on a tripod and heat to boiling for 5 minutes, stirring constantly with a glass rod. Filter into a series of test tubes in a rack and test 2 ml portions of the filtrate in the usual way for the following:

> iron, aluminium, calcium, magnesium,
> potassium, sodium, phosphorus.

Compare the relative abundance of the various elements. These vary somewhat from soil to soil so that it would be instructive to perform the experiment on contrasting soils. Keep the test for comparison with results of Exercise 7.

Exercise 7 – Chemical composition of plants – dried grass

Equipment

10 cm porcelain or silica basin	Filter funnel
Copper wire	Filter paper
Tripod	Glass rod
Bunsen burner	Test tubes in rack

Reagents

2M nitric acid
Usual test reagents for iron, aluminium, calcium, magnesium, potassium, sodium, phosphorus.

Procedure

Almost fill the porcelain basin with dried grass and ignite it on a tripod over a bunsen burner in a fume cupboard. Stir the residue with a copper wire to ensure that ignition is complete and only a white ash remains.

Allow the dish to cool and add 25 ml of 2M nitric acid, heat to boiling and filter into a series of test tubes in a rack. Test 2 ml portions of the filtrate for the following: iron, aluminium, calcium, magnesium, potassium, sodium and phosphorus. Note the relative abundance of the elements in the grass and compare the results with those for the soil from Exercise 6. Account for any differences that are observed.

Exercise 8 – Release of ions by weathering

Most plant nutrient elements come from rocks either directly or indirectly. Therefore they have to be released into the soil solution before they can be taken up by the plant roots. The rate and amount released depends upon the nature of the rock, the acidity of the soil solution and the soil temperature. The interaction of these can be demonstrated experimentally.

Equipment

4 250 ml beakers	4 Tripods with gauzes
Cubes of ice	4 Filter funnels and filter papers
4 Thermometers	Test tubes in rack
4 Bunsen burners	4 Stirring rods

Reagents

Usual test reagents for iron, aluminium, calcium, magnesium, potassium, sodium and phosphorus

0·01 M HCl

0·1   M HCl

1·0   M HCl

Procedure

Weigh out four lots of 20 g of ground granite and place each in a separate 250 ml beaker. Add 100 ml of distilled water to each beaker and quickly adjust the temperature of the water so that each is different and at 1°C, 10°C, 50°C and 100°C respectively. Maintain at these temperatures for 30 minutes stirring regularly.

Filter the suspensions and test 10 ml portions of each filtrate for iron, aluminium, calcium, magnesium, potassium, sodium and phosphorus. Record the relative abundance using a score of plusses. Plot graphs showing the variation in the amount of each ion with temperature. Comment on the results.

Weigh out 20 g of ground granite, basalt, slate and limestone and place each into a separate 250 ml beaker. Add 100 ml of distilled water to each and boil for 30 minutes stirring regularly. Filter the suspensions and test 10 ml portions of each filtrate for the ions mentioned above and record in a similar manner. Draw a histogram to show the differences in the amount of each ion released by the different rocks.

Weigh out four lots of 20 g of basalt and place each in a separate

250 ml beaker. Add 100 ml distilled water to the first beaker, 100 ml 0·01 M HCl to the second, 100 ml 0·1 M HCl to the third, 100 ml 1 M HCl to the fourth. Allow to stand for 30 minutes stirring regularly, then filter the suspension and test 10 ml portions of each filtrate for the ions mentioned above and record the results in a similar manner. Plot graphs showing the variations in the amount of each ion against strength of acid.

The graphs and histograms should show:
1. that rocks differ in the type and amount of ions that they release
2. by increasing the temperature more ions are brought into solution
3. by increasing the acidity more ions are brought into solution.

These laboratory conditions are more drastic than those in nature, but the principles are the same.

Discuss the significance of these results to plant growth.

Exercise 9 – Examination of soil minerals

Apart from a relatively few soils that are composed mainly of organic matter, most are composed of mineral material and it is possible to extract the sand fraction to identify some of the range of minerals present.

Equipment

600 ml beaker	Stirring rod
50 ml cylinder	100 ml porcelain basin
Gauze	Oven set at 105°C
Tripod	Hand lens × 10
Bunsen burner	Pocket magnet

Reagents
20 vols. hydrogen peroxide
2 M hydrochloric acid

Procedure

Place about 50ml soil in a 600ml beaker and add 50ml of 20 vols. hydrogen peroxide. Place the beaker on a gauze resting on a tripod and warm gently with a bunsen to decompose the organic matter present, adding more peroxide if necessary until all of the organic matter has been decomposed. This is indicated by the cessation of effervescence. For many soils no further treatment is necessary but if the grains are coated with iron oxide then add 50 ml of 2M hydrochloric acid and boil for 5–10 minutes. Make a mark on the side of the beaker at 10 cm above the bottom. Fill to the mark with water and stir vigorously, allow to stand for 5 minutes then decant and discard the suspension. Repeat until there is nothing left in suspension. After the final decantation pour off the sand into a porcelain basin and dry at 105°C in an oven.

Place some of the dry sample on to a sheet of paper and spread the sample out so that most of the grains are separated one from another. Examine with a hand lens or binocular microscope if this is available. It will be seen that the grains vary very much in shape and colour but it is usually possible to recognise the common minerals as follows:

Quartz: colourless and irregular, often looking like pieces of broken glass

Feldspars: milky white or pink with irregular shapes

Muscovite: thin transparent or pale coloured flakes

Biotite: dark brown (sometimes brassy) flakes which reflect strongly

Hornblende and

Pyroxenes: dark green to black irregular grains

Magnetite: black and easily removed with a pocket magnet

Calcite: white with irregular shapes, distinguished from feldspars because calcite will effervesce with dilute hydrochloric acid

Exercise 10 – Heat movement in soils

Equipment

2 – 400 ml metal containers e.g. calorimeters
6 thermometers up to 100°C; check that
    they agree with each other
Hot plate
Sand bath
Pipe-clay triangle

Procedure

Almost fill a 400 ml container with soil and place three thermometers at different levels into the soil. One should touch the bottom of the container and the other two should be at 2·5 and 5 cm respectively above the bottom. Record the temperature shown by each thermometer which should be the same and then place the container on a sand bath on a hot plate set at about 100°C. Record the temperature of each thermometer at 5 minute intervals for 30–40 minutes. Remove the container and place it on a pipe-clay triangle and continue to record the temperature for a further 30–40 minutes.

While recording the temperatures almost fill a second container with soil and saturate it by adding small amounts of water at a time to one side of the container. In this way the water flows to the bottom of the container and saturates the soil from the bottom upwards thus expelling all of the air. Repeat the experiment with the second container of wet soil. From the data prepare graphs by plotting time against temperature. Comment on your results and discuss their importance in the field.

Exercise 11 — Estimation of organic matter in soils

The determination of the amount of organic matter in the soil is a fairly complicated technique but in most cases the loss in weight due to ignition corresponds very closely to the amount of organic matter present, except when there is much carbonate present.

Equipment

Silica or porcelain crucible	Bunsen burner
Oven set at 105°C	Copper wire
Tripod	Desiccator
Pipe-clay triangle	Crucible tongs

Procedure

Place the crucible on the triangle resting on the tripod and ignite for 10–15 minutes. Allow the crucible to cool for 2–3 minutes and then place it in the desiccator using crucible tongs. When completely cool accurately weigh the crucible empty, then with 2–3 g of soil. Place the crucible in an oven at 105°C overnight, then cool in a desiccator and re-weigh. Place the crucible on the triangle resting on the tripod. Heat gently with the bunsen burner for about 2–3 minutes then gradually increase the heat to the maximum of the burner. Retain it at the maximum heat for 25–30 minutes, stirring at 10 minute intervals with the copper wire. Allow the crucible to cool for 2–3 minutes and then place it in a desiccator using crucible tongs. When completely cool weigh and calculate the loss in weight as a percentage of the oven dry soil.

In most cases the sample changes colour due to ignition, comment on the reason for this change.

Keep the ignited sample for Exercise 12.

If Exercise 1 has been carried out repeat this exercise for each horizon. Then prepare a graph by plotting depth against loss on ignition. This should illustrate that the properties of horizons vary with depth.

Exercise 12 – Demonstration of the presence of living organisms in soils

Equipment
3 pieces of muslin 10×10 cm
Thread
3 – 250 ml conical flasks with rubber bungs

Reagent
Mixed indicator of cresol red and thymol blue

Procedure

Place about 5 g soil on a piece of muslin then slightly moisten and tie it with thread to form a bag leaving one end of the thread about 10 cm long.

Place 25 ml of the mixed indicator into a 25 ml conical flask and suspend the bag in the flask with a tight fitting stopper (a rubber bung is more suitable than a cork).

Repeat the above using the ignited sample from Exercise 11 and an unignited sample moistened with formaldehyde.

Allow all three to stand in a warm place and note that in the presence of the fresh soil the indicator gradually changes from orange to pale yellow as the solution becomes more acid due to the adsorption of CO_2 evolved from the respiring organisms in the soil. In the presence of the other two, the indicator remains unchanged. What is the reason for this?

Exercise 13 – Demonstration of fungi in soils

Equipment

Cellophane	2 × 100 ml beakers
Pair of scissors	Microscope slide
25 cm clock glass	Microscope
150 ml beaker	Wash bottle with distilled water

Reagents
40 per cent acetic acid in a 100 ml beaker
Blue stain – 0·1 per cent cotton blue in lactophenol

Procedure

Boil a sheet of cellophane in water for 30 minutes to remove the water-proofing lacquer. While boiling place 100 ml of surface soil on to a 25 cm clock glass, moisten gradually and mix until the whole soil has changed colour but there should be no free water present. Transfer the soil to a 150 ml beaker.

Cool the cellophane and while still damp cut the sheet into 2·5 cm squares with a pair of scissors. Place a square at one end of a microscope slide and press it on so that it adheres firmly.

Place the slide into the beaker of soil so that the cellophane is completely buried. Care should be taken to prevent the cellophane being removed when putting the slide in place.

Allow the preparation to stand for about a week and then carefully remove the slide and wash off any adhering soil with a jet of distilled water from a washbottle. Immerse the slide in 40 per cent acetic acid for about 3 minutes, wash again with a jet of distilled water and then immerse in the blue stain for 5 minutes. Remove the slide, wash off the excess stain and examine with a hand lens or microscope. The slide can be replaced in the soil and stained and examined at weekly intervals. This will show a succession of colonisation – fungi, bacteria, mites and other fauna.

Exercise 14 – Extraction of some fauna from fresh soil with the Tullgren funnel

Equipment
15 cm domestic strainer
Desk lamp with 25 watt bulb
Funnel to hold the strainer

Stand to support the funnel and strainer
100 ml beaker
Microscope slides and cover glasses

Reagent
Mixture of 70 per cent ethanol and 5 per cent glycerol
Glycerol

Procedure

Take a freshly collected moist surface sample, crumble it and fill
the strainer to within 1 cm from the top. If the sample contains any
fairly large fauna such as earthworms they should be removed.
Place the strainer into the funnel held in the stand and arrange the
desk lamp so that the bulb is 30 cm from the surface of the soil. Put
50 ml of the ethanol reagent in the 100 ml beaker beneath the stem
of the funnel, switch on the lamp and leave to stand for three to four
days. The lamp will warm the soil so that it gradually heats from the
surface downwards. This causes the soil fauna to move downwards
and to fall out of the soil through the mesh of the strainer and to
collect in the beaker. If the lamp is too close to the soil it will kill
the organisms before they can move downwards.

The organisms can now be mounted on a microscope slide with
glycerol and examined under a microscope. Identifying all the
various organisms may be impossible but it is assumed that you
have a working knowledge of biology.

Exercise 15 – Estimation of earthworm population

In order to estimate the population of earthworms in the soil they
have to be extracted from the soil, counted and weighed. There are a
number of field techniques involving the use of chemicals which
cause worms to come to the surface of the soil from which they can
be collected. Unfortunately these methods usually lead to the death
of the worms and furthermore they usually do not remove the
complete worm population.

It has been found that manual extraction kills very few of the
worms and gives more reliable results.

Equipment

Spade	2 Plastic beakers
Polythene sheet 1 m square	Domestic scales

Procedure

Dig out a 25 cm cube of soil and place it in the centre of the polythene sheet. The soil cube should be removed quickly because earthworms move from the surface very fast when disturbed. Gradually crumble the soil between the fingers and remove all the worms. Note the number of worms that is removed and as each worm is removed wash it in water in one of the beakers and then place it into the other beaker.

After all the worms have been removed weigh them, remix with the soil and replace the mixture back in the hole and wet thoroughly.

Calculate the weight and number of worms per hectare.

If possible repeat the exercise at several different depths and sites in order to demonstrate the variations in earthworm populations.

Exercise 16 – Estimation of texture

Procedure

1. Place a small portion of soil in the palm of the hand, moisten it gradually, manipulate it between each addition of water until the state of maximum plasticity is reached. This is the point at which the soil can be moulded most easily but there is no evidence of free water and the soil does not stick to a polished surface.

2. Press and remould the soil between the finger and thumb several times and note the grittiness, smoothness, stickiness and plasticity of the soil and the degree of "polish" that can be obtained on the surface.

Grittiness is a measure of the amount of sand present. The greater the amount of sand the more the individual particles will rub together causing a gritty feel which can also be heard if held near to the ear.

Smoothness is a measure of the amount of silt. As the silt content increases so does the smoothness or slipperiness which sometimes make the soil feel silky or soapy. This is because silt particles glide or slip one over the other.

Stickiness, plasticity and cohesion are measures of the amount of clay. They become more marked as the clay content increases. The greater the plasticity the easier it is to mould the sample into balls and threads. The surface of clay soils also "takes on a polish" when rubbed gently.

3. Roll the soil round and round between the palms of the hands

to produce a ball. Note the ease with which the ball is produced and its cohesiveness.

4. Roll the ball backwards and forwards between the palms of the hands to produce threads. Note the ease with which threads are produced.

5. If threads are produced, try to form rings.

6. From the results obtained determine the texture of the soil using the key (page 149).

7. The approximate particle size distribution can be determined by examining Fig. 39.

Exercise 17 – Investigation of particle size distribution

Equipment
500 ml volumetric flask with cork or rubber stopper
Stand with clamp

Procedure

Place about 25 ml of soil in a 500 ml volumetric flask, add about 300 ml water and stopper securely. Shake vigorously for about two minutes to break down any aggregates and to disperse the fine particles. Unstopper and almost fill the flask with water, replace the stopper and shake again, invert the flask quickly, secure it in that position with the stand and allow to stand making regular observation. It is seen that a layering develops with the coarsest particles at the bottom and the finest at the top. If the soil was well dispersed it is possible to see four distinct layers. The bottom layer is composed of coarse sand, the individual grains being easily discerned with the unaided eye. Above is the fine sand in which particles are just visible. Then comes the silt which has a homogeneous colour and no particles can be seen with the unaided eye. Finally at the top there is usually a distinct layer of clay which may take 24 hours or more to settle but usually some remains in suspension.

If the soil contains more than about 5 per cent organic matter as determined from Exercise 11 it should be treated with hydrogen peroxide as given in Exercise 9 before putting it into the flask.

Measure the thickness of each layer and calculate each as a percentage of the total, thus obtaining a rough indication of the particle size distribution. Determine the texture from Fig. 39 and compare the results obtained from Exercise 16 – Estimation of texture; they may or may not give similar results.

TABLE III: *Assessment of Soil Texture*

GRITTINESS	SMOOTHNESS	STICKINESS AND PLASTICITY	BALL AND THREAD FORMATION	TEXTURE
Non-gritty to slightly gritty	Not smooth	Extremely sticky and plastic	Extremely cohesive balls and long threads which bend into rings easily. High degree of polish.	Clay
	Moderately smooth and silky	Very sticky and plastic	Very cohesive balls and long threads which bend into rings. High degree of polish.	Silty clay
		Moderately sticky and plastic	Moderately cohesive balls, forms threads which will not bend into rings. Moderate degree of polish	Silty clay loam
	Extremely smooth and silky	Very slightly sticky and plastic	Moderately cohesive balls, forms threads with difficulty that have broken appearance. Slight degree of polish	Silt
	Very smooth and silky	Slightly sticky and plastic	Moderately cohesive balls, forms threads with great difficulty that have broken appearance. No polish	Silt loam
Slightly to moderately gritty	Slightly smooth	Moderately sticky and plastic	Very cohesive balls, forms threads which will bend into rings	Clay loam
Moderately gritty	Not smooth	Very sticky and plastic	Very cohesive balls, forms long threads which bend into rings with difficulty. High degree of polish	Sandy clay
	Not smooth	Moderately sticky and plastic	Moderately cohesive balls, forms long threads which bend into rings with difficulty. Moderate degree of polish	Sandy clay loam
	Slightly smooth	Slightly sticky and plastic	Moderately cohesive balls, form threads with great difficulty	Loam
Very gritty	Not smooth	Not sticky or plastic	Slightly cohesive balls, does not form threads	Sandy loam
Extremely gritty	Not smooth	Not sticky or plastic	Slightly cohesive balls, does not form threads	Loamy sand
		Not sticky or plastic	Non-cohesive balls which collapse easily	Sand

Exercise 18 – Water retaining capacity of soils

Soils have a varying capacity for absorbing and retaining moisture depending upon their texture, structure, and organic matter
content.

Equipment
4 Glass tubes 20 cm long 5 cm diameter
4 Corks with a hole to fit the glass tubes
Clamps for tubes
50 ml cylinder
Cotton wool
4 Stands to hold the tubes
4 – 400 ml beakers

Procedure
Place each cork into one end of each glass tube, put a 1 cm
layer of cotton wool into each tube over the cork. Fill each tube to
within 7 cm of the top with a different one of the soils collected in
Exercise 2. Clamp the tubes in the vertical position, place a beaker
beneath each tube and pour 100 ml of water into each tube and allow
to stand. The water will percolate through each soil at a different
rate and will then drain into the beaker. When drainage is complete
measure the amount of water in each beaker. The difference between
the amount drained through and that added is the amount retained
by the soil.

Exercise 19 – Determination of pH

The pH value of a soil can be determined roughly in the field or
more accurately in the laboratory. In most modern laboratories it is
customary to use a pH meter for this determination. As the construction of pH meters differs very widely and they are not usually
available in elementary laboratories, they will not be discussed.

The simplest method for obtaining an estimate of soil pH is by
using pH papers as described below, or by using one of a variety of
colorimetric kits that are available commercially.

Equipment
Porcelain dish   pH papers

Procedure
Place a strip of pH paper in the bottom of a porcelain dish. Cover

about $\frac{2}{3}$ of the paper with soil which is then gradually moistened until it is saturated and moisture is drawn along part of the exposed length of the paper. The paper will change colour which is compared with the colour range on the packet to determine the soil pH.

Exercise 20 – Determination of the lime requirement

When the pH of a cultivated soil falls below the optimum for crop growth it is usual to add liming material to the soil to raise the pH. In Britain the optimum pH is about 6·5 and the amount of liming material required to produce that pH in various soils can be determined fairly accurately in the laboratory or can be estimated roughly using Table IV (Pizer 1961).

Equipment
5 – 50 ml wide neck bottles with bungs
25 ml pipette
Shaker
pH meter

Reagents
Calcium hydroxide – pure and carbonate free

Procedure

Weigh 10·0 g of soil into each of the five wide neck bottles. Add respectively 0·0, 0·01, 0·02, 0·04 and 0·08 g of calcium hydroxide to the bottles, then add 25 ml distilled water to each bottle, stopper tightly and shake for 16 hours.

Determine the pH of each soil suspension with a pH meter, washing the electrodes between each determination. Prepare a graph by plotting pH (vertical ordinate) against the varying amounts of calcium hydroxide and read off the amount of calcium hydroxide required to raise the pH to 6·5. Convert this by calculation to tonnes per hectare per 15 cm depth of soil of calcium carbonate. This is the lime requirement.

If a shaker is not available, shake the bottles occasionally for 3–4 days. In this case add 2 drops of chloroform to each bottle to prevent the growth of microorganisms. It would be instructive to repeat the experiment with the range of samples collected in Exercise 2 and to compare the results obtained using Table IV.

TABLE IV:

Assessment of the lime requirement of soils

from their pH value, texture and content of organic matter

(tonnes per hectare)

pH value		Loamy sands	Sandy loams	Loams, Silty loams, Silt loams, Organic sandy loams	Clay loams, Organic loams, Silt loams, Peat loams	Clays, Organic clay loams, Peat
	SOIL TYPE					
6·0		20	25	30	40	50
5·5		40	50	65	80	100
5·0		60	75	95	120	150
4·5		80	95	130	160	190
4·0		100	120	160	200	240
3·5		120	145	190	245	290

Exercise 21 — Demonstration of exchange properties

Equipment

3 – 400 ml beakers
3 stirring rods
3 filter funnels and fine filter papers
10 test tubes in rack

Reagents

Distilled water
1M sodium chloride
1M potassium hydrogen phosphate
Usual test reagents for calcium, potassium,
    chloride and phosphorus

Procedure

Place about 25 ml soil into each of the three 400 ml beakers. Add 50 ml of distilled water to the first beaker, 50 ml of sodium chloride solution to the second beaker and 50 ml of potassium hydrogen phosphate solution to the third beaker. Stir each for about 5 minutes and then allow to stand for about 5 minutes until most of the soil has settled. Filter each separately into test tubes and test 5 ml portions for calcium, potassium and chloride in the first two filtrates and phosphorus in the third. Also carry out a similar test for phosphorus in the original solution to compare with the filtrate results. For most soils the calcium and potassium in the filtrate from the sodium chloride treatment should be greater than that of the water treatment. This shows that very little material is readily soluble in water; it shows also that sodium has replaced calcium and potassium on the exchange complex and that they have come into solution. The amounts of chloride in the original liquids and the filtrates should be the same, indicating that most anions are not held in the soil. The phosphorus in the filtrate from the phosphate treatment should be less than in the original solution thus showing that soils can absorb certain anions particularly phosphate.

Appendix III. Projects

Set out below are a few Projects that could be carried out by a single individual or a small class. The details are kept to a minimum for it is assumed that the class will work out the details in collaboration with a supervisor.

Project 1. Measurement of the variations in the moisture content in the surface soil

The amount of moisture in the top 15 cm of the soil varies from season to season and with increasing time following a shower of rain. These variations can be measured very easily by collecting duplicate samples at the following times:
  a) 1 hour, 2 hours, ... 24 hours after a heavy shower of rain.
  b) every day at the same time for 7–10 days in the middle of each season.
  c) every week on the same day at the same time.
Each sample should be $10 \times 10 \times 15$ cm deep, placed into a labelled polythene bag and sealed immediately after it has been collected.

The samples should be taken to the laboratory and weighed in the polythene bags. Then spread them in trays or on polythene sheets and allow to dry. When dry reweigh and calculate the moisture as a percentage of the air dry soil remembering to subtract the weights of the polythene bags.

From the results construct graphs to show; 1) the rate of moisture change in the soil following a rain shower; 2) the seasonal variation in soil moisture; 3) the annual variation in soil moisture. If possible conduct the project on the upper slopes of a low hill as well as in the valley. The two sets of data should be very different.

Project 2. Demonstration of soil variability

Soils may vary considerably over relatively short distances. This can be demonstrated by measuring a number of properties such as pH and loss on ignition.

Lay out a 100 m grid with 10 m intersections on a site which seems to show some variation. Number each intersection and collect a 10 × 10 × 15 cm sample at each intersection. Place each sample in a labelled polythene bag, take to the laboratory and prepare in the usual manner (Exercise 4). Determine the pH of each sample with a pH meter if possible. Also determine the loss on ignition on an air dried portion of the samples (see Exercise 11).

Prepare two 2 cm grids on a piece of paper and number the intersections in a similar manner to the field layout. Put the results for the pH on one grid and the results for the loss on ignition on the other. Note the trends in the variations and comment on the results.

Project 3. Soil examination at different places in the landscape

Choose sites beneath an oak wood, a beech wood, a coniferous forest, a heath and a grassland or three other contrasting types of vegetation. Dig soil pits at each site and describe the soils. Collect samples from each horizon (see Exercise 1) and determine the pH and loss on ignition of each sample. From the results prepare two graphs for each soil by plotting depth against pH and against loss on ignition. Comment on the results.

Project 4. Variations in stone shape

Many soils contain stones which can be a useful guide to the origin of the material (see page 56). During the course of the practical work collect stones 5–10 cm in size having as many shapes as possible. Make drawings of the most distinctive shapes and account for their origin. The more enterprising participants might want to take photographs of the specimens.

Project 5. Measurement of evapotranspiration and evaporation

Equipment

Rake	Trays for drying samples
Spade	Rain gauge
Polythene bags, rubber bands and labels	Polythene bucket with wire mesh cover
Trowel	1 litre graduated measuring
Balance	cylinder

Procedure

Choose a site on a freely draining soil that has a thick growth of vegetation which may be a crop, grassland or even weeds. Mark out two adjacent plots each 3m square. Carefully remove all the vegetation from one of the plots and lightly cultivate it with a rake to a depth of 2–3 cm. At the borders of the cleared plot continuously insert a spade in order to sever any roots from plants on the adjacent sites.

Very quickly dig a small hole $25 \times 25 \times 50$ cm deep in one corner of one of the plots about 50 cm from the two nearest sides. Also very quickly collect directly into a polythene bag a sample of soil of about 200 g from each of the following depths: 5, 10, 20 and 50 cm. Quickly seal and label each polythene bag. Refill the holes carefully and replace the turf in the case of the vegetated plot. It is essential to work quickly in order to reduce water losses due to evaporation from the freshly exposed soil surfaces.

Repeat the procedure for the second plot. Transport the samples to the laboratory. Weigh the whole and then place the samples in trays to dry and then reweigh. Calculate the percentage of moisture in each sample remembering to subtract the weight of the polythene bags. Repeat the determinations at monthly intervals. If it is planned to run the experiment for a long period the size of the plots will have to be larger so that more small holes can be dug within the plot area.

Running in parallel with the above it is necessary to measure the precipitation and also the evaporation from a free water surface. Beside the plots place a rain gauge and a plastic bucket containing a measured volume of water (4–5 litres). It is essential to place a wire mesh over the bucket otherwise various animals will drink the water.

Read the rain gauge and measure the water in the bucket at regular intervals (weekly). If from time to time the bucket appears to be too empty or too full add or remove a measured volume. From the results calculate the amount of precipitation and evaporation on a weekly and monthly basis.

Prepare several graphs and histograms showing the variations in the moisture contents in the soil both with depth and time for the individual plots and between plots. Wherever possible relate the amount and intensity of rainfall to the soil moisture patterns. Discuss fully the results.

Project 6. Preparation of miniature monoliths

Professional soil scientists often collect monoliths of soil for exhibition and demonstration. This is a very laborious and expensive operation but it is possible to prepare small monoliths from the samples collected in Exercise 1, or Project 3.

Make a scale drawing of the horizons in the soil profile on a piece of cardboard 10 cm wide and 40 cm long. Starting at the top apply a quick setting glue to the space for the uppermost horizon then sprinkle onto the glue some of the soil sample. Allow the glue to dry then apply another layer of glue over the soil and again sprinkle some soil on to the wet glue. Repeat the operation three or four times until a realistic appearance of soil develops. Then do the remaining horizons, taking care to produce the correct boundaries between the horizons.

Project 7. Demonstration of soil colour variations

The colour of soils varies considerably from place to place and during the course of the practical work it would be very instructive to collect soils of different colours. It will only be necessary to collect small samples which are air dried and sieved as given in Exercise 4. These samples can then be used for exhibition purposes in a variety of ways. They can be put into small bottles or they can be stuck onto circular areas on large sheets of cardboard in a similar manner to the miniature monoliths technique described in Project 6.

Project 8. Investigation into the changes in land use at a given site

In countries such as Britain some areas have been mapped several times over the last two centuries. Choose such an area and with the aid of the maps and any other information prepare an account of the changes in land use that have occurred.

Project 9. Exhibition

If all of the exercises and projects have been completed it should be possible to prepare an exhibition of the results and specimens.

Glossary

ACCELERATED EROSION: An increased rate of erosion caused by man.

ACIDITY: The hydrogen ion activity in the soil solution expressed as a pH value.

ACID ROCK: An igneous rock that contains more than 60 per cent silica and free quartz.

ACID SOIL: A soil with pH < 7.0.

ACTINOMYCETES: A group of organisms intermediate between the bacteria and the true fungi, mainly resembling the latter because they usually produce branched mycelium.

ADSORPTION: The attachment of a particle, ion or molecule to a surface. Calcium is adsorbed on to the surface of clay or humus.

ADSORPTION COMPLEX: The various substances in the soil that are capable of adsorption, these are mainly the clay and humus.

AEOLIAN: Pertaining to or formed by wind action.

AEOLIAN DEPOSITS: Fine sediments transported and deposited by wind; they include loess, dunes, desert sand and some volcanic ash.

AERATION: The process by which atmospheric air enters the soil. The rate and amount of aeration depends upon the size and continuity of the pore spaces and the degree of water logging.

AEROBIC: Conditions having a continuous supply of molecular oxygen.

AEROBIC ORGANISM: Organisms living or becoming active in the presence of molecular oxygen.

AGGREGATE: A cluster of soil particles forming a ped, cf. Fragment.

AGGREGATION: The process by which particles coalesce to form aggregates.

ALGAE: Unicellular or multicellular plants containing chlorophyll. They are aquatic or occur in damp situations and include most seaweeds.

ALKALINE SOIL: A soil with pH > 7.0.

ALLUVIAL SOIL: A general term for those soils developed on fairly recent alluvium.

ALLUVIUM: A sediment deposited by streams and varying widely in particle size. The stones and boulders when present are usually rounded or sub-rounded. Some of the most fertile soils are derived from alluvium of medium or fine texture.

AMINO ACID: An organic compound containing both the amino (NH_2) and carboxyl (COOH) groups. Amino acid molecules combine to form proteins, therefore they are a fundamental constituent of living matter. They are synthesised by autotrophic organisms, principally green plants.

AMMONIFICATION: The production of ammonia by microorganisms through the decomposition of organic matter.

ANAEROBIC: Conditions that are free of molecular oxygen. In soils this is usually caused by excessive wetness.

ANAEROBIC ORGANISM: One that lives in an environment without molecular oxygen.

ANION: An ion having a negative charge.

ARTHROPOD: A member of the phylum arthropoda which is the largest in the animal kingdom. It includes, insects, spiders, centipedes, crabs, etc.

AUGER: See Soil Auger.

AUTOTROPHIC ORGANISMS: Organisms that utilise carbon dioxide as a source of carbon and obtain their energy from the sun or by oxidising inorganic substances such as sulphur, hydrogen, ammonium, and nitrate salts. The former include the higher plants and algae and the latter various bacteria, cf. Heterotrophic.

AVAILABLE ELEMENTS: The elements in the soil solution that can readily be taken up by plant roots.

AVAILABLE NUTRIENTS: See AVAILABLE ELEMENTS.

AVAILABLE WATER CAPACITY: The weight percentage of water which a soil can store in a form available to plants. It is equal to the moisture content at field capacity minus that at the wilting point.

BACTERIA: Unicellular or multicellular microscopic organisms. They occur everywhere and in very large numbers in favourable habitats such as soil and sour milk where they number many millions per gram.

BASALT: A fine-grained igneous rock-forming lava flows or minor intrusions. It is composed of plagioclase, augite and magnetite; olivine may be present.

BASE SATURATION: The extent to which the exchange sites of a material are occupied by exchangeable basic cations; expressed as % of the cation exchange capacity.

BASIC ROCK: An igneous rock that contains less than 55 per cent silica.

BIENNIAL: A plant that completes its life cycle in two years.

BIOMASS: a) The weight of a given organism in a volume of soil that is one m^2 at the surface and extending down to the lower limit of the organisms penetration.
b) The weight of organisms in a given area or volume.

BOG IRON ORE: A ferruginous deposit in bogs and swamps formed by oxidizing algae, bacteria or the atmosphere on iron in solution.

BOULDER CLAY: See TILL.

BS: an abbreviation for Base Saturation.

BUFFER: A substance that prevents a rapid change in pH when acids or alkalis are added to the soil, these include the clay, humus and carbonates.

BULK DENSITY: Mass per unit volume of undisturbed soil, dried to constant weight at 105°C. Usually expressed as g/cm^3.

CALCAREOUS SOIL: A soil that contains enough calcium carbonate so that it effervesces when treated with hydrochloric acid.

CALCITE: Crystalline calcium carbonate, $CaCO_3$. Crystallises in the hexagonal system, the main types of crystals in soils being dog-tooth, prismatic, fibrous, nodular, granular and compact.

CAPILLARITY: The process by which moisture moves in any direction through the fine pore spaces and as films around particles.

CAPILLARY MOISTURE: That amount of water that is capable of movement after the soil has drained. It is held by adhesion and surface tension as films around particles and in the finer pore spaces.

CATION: An ion having a positive electrical charge.

CATION EXCHANGE: The exchange between cations in solution and cations held on the exchange sites of minerals and organic matter.

CATION EXCHANGE CAPACITY: The total potential of soils for adsorbing cations, expressed in milligram equivalents per 100 g of soil. Determined values depend somewhat upon the method employed.

CEMENTED: Massive and either hard or brittle depending upon the content of cementing substances such as calcium carbonate, silica, oxides of iron and aluminium, or humus.

CHALK: The term refers to either (a) soft white limestone which consists of very pure calcium carbonate and leaves little residue when treated with hydrochloric acid, sometimes consists largely of the remains of foraminifera, echinoderms, molluscs and other marine organisms, or (b) the upper or final member of the Cretaceous System.

CHLOROSIS: The formation of pale green or yellow leaves in plants resulting from the failure of chlorophyll to develop. It is often caused by a deficiency in an essential element.

CLAY: Either 1. Mineral material $<2\mu m$
        2. A class of texture
        3. Silicate clay minerals.

CLAY MINERAL: Crystalline or amorphous mineral material, $<2\mu m$ in diameter.

CLOD: A mass of soil produced by disturbance.

CONIFEROUS FOREST: A forest consisting predominantly of cone-bearing trees with needle-shaped leaves: usually evergreen but some are deciduous, for example the larch forests (*Larix dehurica*) of central Siberia. Their greatest extent is in the wide belt across northern Canada and northern Eurasia. Coniferous forests produce soft wood which has a large number of industrial applications including paper making.

COLLOID: The inorganic and organic material with very fine particle size and therefore high surface area which usually exhibits exchange properties.

COLLUVIUM: Soil materials with or without rock fragments that accumulate at the base of steep slopes by gravitational action.

COMPOST: Plant and animal residues that are arranged into piles and allowed to decompose, sometimes soil or mineral fertilizers may be added.

CONCEPT: General notion.

CONCRETION: Small hard local concentrations of material such as calcite, gypsum, iron oxide or aluminium oxide. Usually spherical or subspherical but may be irregular in shape.

CONGLOMERATE: A sedimentary rock composed mainly of rounded boulders.

CONSISTENCE: The resistance of the soil to deformation or rupture as determined by the degree of cohesion or adhesion of the soil particles to each other.

CONSOLIDATED: A term that usually refers to compacted or cemented rocks.

CREEP: Slow movement of masses of soil down slopes that are usually steep. The process takes place in response to gravity facilitated by saturation with water.

CROTOVINA: An animal burrow which has been filled with material from another horizon.

CUTANS: Coatings or deposits of material on the surfaces of peds, stones, etc. A common type is the clay cutan caused by translocation and deposition of clay particles on ped surfaces.

DECIDUOUS FOREST: A forest composed of trees that shed their leaves at some season of the year. In tropical areas the trees lose their leaves during the hot season in order to conserve moisture. Deciduous forests of the cool areas shed their leaves during the autumn to protect themselves against the cold and frost of winter. Deciduous forests produce valuable hardwood timber such as teak and mahogany from the tropics, oak and beech come from the cooler areas.

DEFLOCCULATE: To separate or disperse particles of clay dimensions from a flocculated condition.

DELTA: A roughly triangular area at the mouth of a river composed of river transported sediment.

DENITRIFICATION: The biological reduction of nitrate to ammonia, molecular nitrogen or to the oxides of nitrogen, resulting in the loss of nitrogen into the atmosphere and therefore undesirable in agriculture.

DENUDATION: Sculpturing of the surface of the land by weathering and erosion; levelling mountains and hills to flat or gently undulating plains.

DEVONIAN: A period of geological time extending from 320–280 million years B.P.

DISPERSION: The process whereby the structure or aggregation of the soil is destroyed so that each particle is separate and behaves as a unit.

DRIFT: A generic term for superficial deposits including till (boulder-clay), outwash gravel and sand, alluvium, solifluction deposits and loess.

DRUMLIN: A small hill, composed of glacial drift with hog-back outline, oval plan, and long axis oriented in the direction of ice movement. Drumlins usually occur in groups, forming what is known as basket of eggs topography.

DRY-FARMING: A method of farming in arid and semi-arid areas without using irrigation, the land being treated so as to conserve moisture. The technique consists of cultivating a given area in alternate years allowing moisture to be stored in the fallow year. Moisture losses are reduced by producing a mulch and removing weeds. In Siberia, where melting snow provides much of the moisture for spring crops, the soil is ploughed in the autumn providing furrows in which snow can collect, preventing it from being blown away and evaporated by strong winds. Usually alternate narrow strips are cultivated in an attempt to reduce erosion in the fallow year. Dry farming methods are employed in the drier regions of India, U.S.S.R., Canada and Australia.

DUNES, SAND DUNES: Ridges or small hills of sand which have been piled up by wind action on sea coasts, in deserts and elsewhere. Barkhans are isolated dunes with characteristic crescentic forms.

ECOLOGY: The study of the interrelationships between individual organisms and between organisms and their environment.

ECOSYSTEM: A group of organisms interacting among themselves and with their environment.

EDAPHOLOGY: The study of the relationships between soil and organisms including the use of the land by man.

ELUVIAL HORIZON: A horizon from which material has been removed either in solution or suspension.

ELUVIATION: Removal of material from the upper horizon in solution or suspension.

EQUATORIAL FOREST OR TROPICAL RAIN FOREST: A dense, luxuriant, evergreen forest of hot, wet, equatorial regions containing many trees of tremendous heights, largely covered with lianas and epiphytes. Individual species of trees are infrequent but they include such valuable tropical hardwoods as mahogany, ebony and rubber. Typical equatorial forests occur in the Zaire and Amazon basins and southeastern Asia.

EROSION: The removal of material from the surface of the land by weathering, running water, moving ice, wind and mass movement.

ESKER: A long narrow ridge, chiefly of gravel and sand, formed by a melting glacier or ice sheet.

EVAPOTRANSPIRATION: The combined processes of evaporation and transpiration.

EXCESSIVELY DRAINED: A soil that loses water very rapidly because of rapid percolation.

EXCHANGEABLE CATION: A cation such as calcium that is adsorbed onto a surface, usually clay or humus and is capable of being easily replaced by another cation such as potassium. Exchangeable cations are readily available to plants.

EXFOLIATION: A weathering process during which thin layers of rock peel off from the surface. This is caused by heating of the rock surface during the day and cooling at night leading to alternate expansion and contraction. This process is sometimes termed 'onion skin weathering'.

FEN PEAT: Peat that is neutral to alkaline due to the presence of calcium carbonate.

FERTILIZER: A material that is added to the soil to supply one or more plant nutrients in a readily available form.

FIELD CAPACITY OR FIELD MOISTURE CAPACITY: The total amount of water remaining in a freely drained soil after the excess has flowed into the underlying unsaturated soil. It is expressed as a percentage of the oven-dry soil.

FINE TEXTURE: Containing > 35 per cent clay.

FLOOD PLAIN: The land adjacent to a stream, built of alluvium and subject to repeated flooding.

FRAGMENT: A small mass of soil produced by disturbance.

FREELY DRAINED: A soil that allows water to percolate freely.

FRIABLE: A term applied to soils that when either wet or dry crumble easily between the fingers.

FUNGI: Simple plants that lack chlorophyll and composed of cellular filamentous growth known as hyphae. Many fungi are microscopic but their fruiting bodies, viz. mushrooms and puffballs are quite large.

GASTROPOD: A member of the Gastropoda class of molluscs which includes snails and slugs.

GEOMORPHOLOGY: The study of the origin of physical features of the earth, as they are related to geological structure and denudation.

GILGAI: A distinctive microrelief of knolls and basins that develops on clay soils that exhibit a considerable amount of expansion and contraction in response to wetting and drying.

GLACIAL DRIFT: Material transported by glaciers and deposited, directly from the ice or from the melt water.

GLACIER: A large mass of ice that moves slowly over the surface of the ground or down a valley. They originate in snow fields and terminate at lower elevations in a warmer environment where they melt.

GLEYING: The reduction of iron in an anaerobic environment leading to the formation of grey or blue colours.

GRANITE: An igneous rock that contains quartz, feldspars and varying amounts of biotite and muscovite.

GRAVITATIONAL WATER: The water that flows freely through soils in response to gravity.

GROUNDWATER-TABLE: The upper limit of the ground water.

GULLY: A shallow steep-sided valley that may occur naturally or be formed by accelerated erosion.

GULLY EROSION: A form of catastrophic erosion that forms gullies.

HALOPHYTE: A plant capable of growing in salty soil; i.e. a salt tolerant plant.

HARDPAN: A horizon cemented with organic matter, silica, sesquioxides, or calcium carbonate. Hardness or rigidity is maintained when wet or dry and samples do not slake in water.

HEAVY SOIL (Obsolete): A soil that has a high content of clay and is difficult to cultivate.

HETEROTROPHIC ORGANISMS: Those that derive their energy by decomposing organic compounds, cf. AUTOTROPHIC.

HOLOCENE PERIOD: The period extending from 10,000–0 years B.P.

HORIZON: Relatively uniform material that extends laterally, continuously or discontinuously throught the pedounit; runs approximately parallel to the surface of the ground and differs from the related horizons in many chemical, physical and biological properties.

HUMIFICATION: The decomposition of organic matter leading to the formation of humus.

HUMUS: The well-decomposed, relatively stable part of the organic matter found in aerobic soils.

HYDRATION: The process whereby a substance takes up water.

HYDROLOGIC CONDUCTIVITY: The rate at which water will move through soil.

HYDROLOGIC CYCLE: Disposal of precipitation from the time it reaches the soil surface until it re-enters the atmosphere by evapotranspiration to serve again as a source of precipitation.

HYDROLYSIS: In soils it is the process whereby hydrogen ions are exchanged for cations such as sodium, potassium, calcium and magnesium.

HYDROMORPHIC SOIL: Soils developed in the presence of excess water.

HYGROSCOPIC WATER: Water that is adsorbed on to a surface from the atmosphere.

IGNEOUS ROCK: A rock formed by the cooling of molten magma including basalt and granite.

ILLUVIAL HORIZON: A horizon that receives material in solution or suspension from some other part of the soil.

ILLUVIATION: The process of movement of material from one horizon and its deposition in another horizon of the same soil; usually from an upper horizon to a middle or lower horizon in the pedounit. Movement can also take place laterally.

IMPEDED DRAINAGE: Restriction of the downward movement of water by gravity.

IMPERFECTLY DRAINED: A soil that shows a small amount of reduction of iron due to short periods of waterlogging.

INFILTRATION: The process whereby water enters the soil through the surface.

INTERGRADE: A soil which contains the properties of two distinctive and genetically different soils.

KROTOVINA: see CROTOVINA.

LACUSTRINE: Pertaining to lakes.

LACUSTRINE DEPOSIT: Materials deposited by lake waters.

LANDSLIDE OR LANDSLIP: The movement down the slope of a large mass of soil or rocks from a mountain or cliff. Often occurs after torrential rain which soaks into the soil making it heavier and more mobile. Earthquakes and the undermining action of the sea are also causative agents.

LATTICE STRUCTURE: The orderly arrangement of atoms in crystalline material.

LIGHT SOIL: (obsolete) A soil which has a coarse texture and is easily cultivated.

LIME: Compounds of calcium used to correct the acidity in soils.

LITTER: The freshly fallen plant material occurring on the surface of the ground.

LODGING: The collapse of top-heavy plants, particularly grain crops because of excessive growth or beating by rain.

LOESS: An aeolian deposit composed mainly of silt which originated in arid regions, from glacial outwash or from alluvium. It is usually of yellowish brown colour and has a widely varying calcium carbonate content. In the U.S.S.R., loess is regarded as having been deposited by water.

MACROELEMENT: Elements such as nitrogen that are needed in large amounts for plant growth.

MANURE: Animal excreta with or without a mixture of bedding or litter.

MERISTEM: The region of active cell-division in plants, it is the tips of stems and roots in most plants. The cells so formed then become modified to form the various tissues such as the epidermis and cortex.

MESOFAUNA: Small organisms such as worms and insects.

MICROCLIMATE: The climate of a very small region.

MICROELEMENT: Those elements that are essential for plant growth but are required only in very small amounts.

MICROFAUNA: The small animals that can only be seen with a microscope; they include protozoa, nematodes, etc.

MICROFLORA: The small plants that can only be seen with a microscope; they include algae, fungi, bacteria, etc.

MICROORGANISM: The members of the microflora and microfauna that can only be seen with a microscope.

MICRORELIEF: Small differences in relief that have differences in elevation up to about 2m.

MILLIEQUIVALENT: A thousandth of an equivalent weight.

MINERAL SOIL: A soil that is composed predominantly of mineral material cf. ORGANIC SOIL.

MITES: Very small members of the arachnid which includes spiders; they occur in large numbers in many organic surface soils.

MORAINE: Any type of constructional topographic form consisting of till and resulting from glacial deposition.

MOTTLING: Patches or spots of different colours.

MULCH: A loose surface horizon that forms naturally or may be produced by cultivation and consists of either inorganic or organic materials.

NEUTRAL SOIL: A soil with pH value around 7.

NITRIFICATION: The oxidation of ammonia to nitrite and nitrite to nitrate by microorganisms.

NITROGEN FIXATION: The transformation of elemental nitrogen to an organic form by microorganisms.

NON-SILICATE: Rock forming minerals that do not contain silicon.

ONION SKIN WEATHERING: See EXFOLIATION.

ORGANIC SOIL: A soil that is composed predominantly of organic matter, usually refers to peat.

PANS: Soil horizons that are strongly compacted, cemented or have a high content of clay.

PARENT MATERIAL: The original state of the soil. The relatively unaltered lower material in soils is often similar to the material in which the horizons above have formed.

PEAT: An accumulation of dead plant material often forming a layer many metres deep. It is only slightly decomposed due to being completely waterlogged.

PED: A single individual naturally occurring soil aggregate such as a granule or prism cf. CLOD or FRAGMENT.

PEDOGENESIS: The natural process of soil formation.

PEDOLOGY: The study of soils as naturally occurring phenomena taking into account their composition distribution and method of formation.

PEDOUNIT: A selected column of soil containing sufficient material in each horizon for adequate laboratory characterisation.

PERCHED WATER: An accumulation of water within the soil due to an impermeable layer such as a pan or a high content of clay.

PERCHED WATER-TABLE: The upper limit of perched water. See Perched Water.

PERCOLATION: (soil water) The downward or lateral movement of water through soil.

PERENNIAL: A plant that continues to grow from year to year.

PERMAFROST: Permanently frozen subsoil.

PERMANENT WILTING POINT: See WILTING POINT.

PERMEABILITY: The ease with which air, water, or plant roots penetrate into or pass through a specific horizon.

pH: The negative logarithm of the hydrogen ion concentration of a solution. It is the quantitative expression of the acidity and alkalinity of a solution and has a scale that ranges from 0 to 14. pH 7 is neutral <7 is acid and >7 is alkaline.

pH SOIL: The negative logarithm of the hydrogen ion concentration of a soil. The degree of acidity (or alkalinity) of a soil expressed in terms of the pH scale, from 2 to 10.

PHYSICAL WEATHERING: The comminution of rocks into smaller fragments by physical forces such as frost action and exfoliation.

PHYSIOLOGICAL DROUGHT: A temporary daytime state of drought in plants due to the losses of water by transpiration being more rapid than uptake by roots even although the soil may have an adequate supply. Such plants usually recover during the night.

PLAGIOCLIMAX: A plant community which is maintained by continuous human activity of a specific nature such as burning or grazing.

PLASTIC: A moist or wet soil that can be moulded without rupture.

PLATY: Soil aggregates that are horizontally elongated.

PLEISTOCENE PERIOD: The period following the Pliocene period, extending from 2,000,000–10,000 years BP. In Europe and North America, there is evidence of four or five periods of intense cold during this period, when large areas of the land surface were covered by ice – glacial periods. During the interglacial periods the climate ameliorated and the glaciers retreated.

POLDER: A term used in Holland for an area of land reclaimed from the sea or a lake. A dyke is constructed around the area which is then drained by pumping the water out. Polders form valuable agricultural land or pasture land for cattle.

POLYGENIC SOIL: A soil that has been formed by two or more different and contrasting processes so that all the horizons are not genetically related.

POORLY DRAINED: A soil that remains very wet or waterlogged for long periods of the year and as a result develops a mottled pattern of greys and browns.

PORE: A discrete volume of soil atmosphere completely surrounded by soil (cf. PORE SPACE).

PORE SPACE: The continuous and interconnecting spaces in soils.

POROSITY: The volume of the soil mass occupied by pores and pore space.

PRIMARY MINERAL: 1. A mineral such as feldspar or a mica which occurs or occurred originally in an igneous rock.

2. Any mineral which occurs in the parent material of the soil.

PROFILE: A vertical section through a soil from the surface into the relatively unaltered material.

PUDDLE: To destroy the structure of the surface soil by physical methods such as the impact of rain drops, poor cultivation with implements and trampling by animals.

QUATERNARY ERA: The period of geological time following the Tertiary Era, it includes the Pleistocene and Holocene periods and extends from 2,000,000–0 years B.P.

RAINFALL INTERCEPTION: The interception and accumulation of rainfall by the foliage and branches of vegetation.

RAIN SPLASH: The redistribution of soil particles on the surface by the impact of rain drops. On slopes this can cause a large amount of erosion.

RAIN SPLASH EROSION: See RAIN SPLASH.

RHIZOSPHERE: The soil close to plant roots where there is usually an abundant and specific microbiological population.

RILL: A small intermittent water course with steep sides.

RILL EROSION: The formation of rills as a consequence of poor cultivation.

REGOLITH: The unconsolidated mantle of weathered rock, soil and superficial deposits overlying solid rock.

SALINE SOIL: A soil containing enough soluble salts to reduce its fertility.

SALINISATION: The process of accumulation of salts in soil.

SAND: Mineral or rock fragments that range in diameter from 2–0·02 mm in the international sytem or 2–0·05 mm in the U.S.D.A. system.

SECONDARY MINERAL: Those minerals that form from the material released by weathering. The main secondary minerals are the clays and oxides.

SEDIMENTARY ROCK: A rock composed of sediments with varying degrees of consolidation. The main sedimentary rocks include sandstones, shales, conglomerates and some limestones.

SELF-MULCHING SOIL: A soil with a naturally formed well aggregated surface which does not crust and seal under the impact of raindrops.

SESQUIOXIDES: Usually refers to the combined amorphous oxides of iron and aluminium.

SHEET EROSION: The gradual and uniform removal of the surface soil by water without forming any rills or gullies.

SILICATES: Rock forming minerals that contain silicon.

SILT: Mineral particles that range in diameter from 0·02–0·002 mm in the international system or 0·05–0·002 mm in the U.S.D.A. system.

SLICKENSIDE: The polished surface that forms when two peds rub against each other when some soils expand in response to wetting.

SLICKSPOT: Small areas of surface soil that are slick when wet because of alkalinity or high exchangeable sodium.

SOIL: The natural space-time continuum occurring at the surface of the earth and supporting plant life.

SOIL AUGER: A tool used for boring into the soil and withdrawing small samples for field or laboratory examination.

SOIL HORIZON: See HORIZON.

SOIL MONOLITH: A vertical section through the soil preserved with resin and mounted for display.

SOIL PROFILE: A section of two dimensions extending vertically from the earth's surface so as to expose all the soil horizons and a part of the relatively unaltered underlying material.

SOIL SURVEY: The systematic examination and mapping of soil.

SOLIFLUCTION: Slow flow of material on sloping ground, characteristic of, though not confined to, regions subjected to alternate periods of freezing and thawing.

SOLUM: The part of the soil above the relatively unaltered material.

SPRINGTAILS: Very small insects that live in the surface soil.

STRIP CROPPING: The practice of growing crops in strips along the contour in an attempt of reduce run off, thereby preventing erosion or conserving moisture.

STRUCTURE: The spatial distribution and total organisation of the soil system as expressed by the degree and type of aggregation and the nature and distribution of pores and pore space.

SYMBIOSIS: Two organisms that live together for their mutual benefit. Fungus and alga that form a lichen or nitrogen fixing bacteria living in roots are examples of symbiosis. The individual organisms are called symbionts.

TECTONIC: Rock structures produced by movements in the earth's crust.

TERRACE: A broad surface running along the contour. It can be a natural phenomenon or specially constructed to intercept run off – thereby preventing erosion and conserving moisture. Sometimes they are built to provide adequate rooting depth for plants.

TERTIARY PERIOD: The period of time extending from 75,000,000–2,000,000 years B.P.

TILE DRAIN: Short lengths of concrete or pottery pipes placed end to end at a suitable depth and spacing in the soil to collect water from the soil and lead it to an outlet.

TILL: An unstratified or crudely stratified glacial deposit consisting of a stiff matrix of fine rock fragments and old soil containing subangular stones of various sizes and composition, many of which may be striated (scratched). It forms a mantle from less than 1m to over 100m in thickness covering areas which carried an ice-sheet or glaciers during the Pleistocene and Holocene periods.

TILL PLAIN: A level or undulating land surface covered by glacial till.

TILTH: The physical state of the soil that determines its suitability for plant growth taking into account texture, structure, consistence and pore space. It is a subjective estimation and is judged by experience.

TOPOSEQUENCE: A sequence of soils whose properties are determined by their particular topographic situation.

TOXIC SUBSTANCE: A substance that is present in the soil or the above ground atmosphere that inhibits the growth of plants and ultimately may cause their death.

TRANSLOCATION: Migration of material in solution or suspension from one horizon to another.

TRIASSIC: A period of geological time extending from 190,000,000–150,000,000 years B.P.

TROPICAL RAIN FOREST: See EQUATORIAL FOREST.

UNAVAILABLE NUTRIENTS: Plant nutrients that are present in the soil but cannot be taken up by the roots because they have not been released from the rock or minerals by weathering or from organic matter by decomposition.

UNAVAILABLE WATER: Water that is present in the soil but cannot be taken up by plant roots because it is strongly adsorbed onto the surface of particles.

UNCONSOLIDATED: Sediments that are loose and not hardened.

VENTIFACT: A pebble facetted or moulded by wind action, usually forms in polar and desert areas. The flat facets meet at sharp angles.

VERY POORLY DRAINED: A soil that remains wet and waterlogged for most of the year so that most of the horizons are blue, olive or grey due to the reducing conditions.

VOLCANIC ASH (Volcanic Dust): Fine particles of lava ejected during a volcanic eruption. Sometimes the particles are shot high into the atmosphere and carried long distances by the wind.

WATERLOGGED: Saturated with water.

WATER-TABLE (Ground): The upper limit in the soil or underlying material permanently saturated with water.

WATER-TABLE PERCHED: See PERCHED WATER-TABLE.

WEATHERING: All the physical, chemical and biological processes that cause the disintegration of rocks at or near the surface.

WILTING POINT: The percentage by weight of water remaining in the soil when the plants wilt permanently.

References

BIBBY, J. S. & MACKNEY, D. 1969. Land use capability classification, Tech. Mono. No. 1. Soil Survey of Great Britain. pp. 27

BRIDGES, E. M. 1970. *World Soils*. Cambridge Univ. Press. pp. 89

CLARKE, G. R. 1971. *The Study of the Soil in the field*. Clarendon Press, Oxford.

FITZPATRICK, E. A., 1971. *Pedology – A systematic approach to soil science*. Oliver and Boyd, Edinburgh. pp. 306

GREENWOOD, D. J. 1970. Soil aeration and plant growth. *Prog. App. Chem.* **55**, 423–431.

HUXLEY, J. G. 1943. *TVA adventure in planning*. The Architectural Press, London. pp. 142.

JENNY, H. 1941. *Factors of Soil Formation*. McGraw-Hill Book Co. Inc., pp. 281

KUBIËNA, W. L. 1953. *The Soils of Europe*. Murby, London, pp. 317

MILLAR, C. E., TURK, L. M. & FOTH, H. D. 1965. *Fundamentals of Soil Science*, John Wiley & Sons Inc., pp. 491

PIZER, N. H. 1961. *The practical application of knowledge of soils*. Ministry of Agriculture, Fisheries and Food. Agricultural Land Services, Technical Report No. 8. pp. 15–20

RUSSELL, SIR E. J. 1957. *The World of the Soil*. (New Naturalist Series) Collins, London. pp. 237

—1961. *Soil Conditions and Plant Growth*. Longmans, London. Ninth edition. pp. 688

YAALON, D. H. 1960. Some implications of fundamental concepts of pedology in soil classification. *Trans. 7th Inter. Cong. Soil Sci.*, **IV**, 119–123

See also *Memoirs of Soil Survey of England and Wales*. Harpenden. *Memoirs of Soil Survey of Scotland*. Macaulay Institute, Aberdeen.

Index